The Bends

# The Bends

Compressed Air

in the History of

Science, Diving,

and Engineering

*John L. Phillips, M.D.*

Yale University Press    New Haven and London

Designed by Sonia L. Scanlon.
Set in Bauer Bodoni type by G & S Typesetters, Austin,
Texas.
Printed in the United States of America by BookCrafters,
Inc., Chelsea, Michigan.

Library of Congress Cataloging-in-Publication Data
Phillips, John L., 1965–
    The bends : compressed air in the history of science,
    diving, and engineering / John L. Phillips
        p.   cm.
    Includes bibliographical references and index.
    ISBN 0–300–07125–6 (cloth : alk. paper)
    1. Compressed air.   I. Title.
TJ985.P46   1998
621.5′1—dc21      97-35206
                  CIP

A catalogue record for this book is available from the
British Library.

The paper in this book meets the guidelines for perma-
nence and durability of the Committee on Production
Guidelines for Book Longevity of the Council on Library
Resources.

10 9 8 7 6 5 4 3 2 1

# Contents

# Preface

When the Brooklyn Bridge celebrated its centennial, there was great interest in the history of its construction. As David McCullough recounts in *The Great Bridge*, the first victims of decompression sickness (or the bends) were not scuba divers of the 1940s, but miners and tunnel builders a hundred years earlier. I appreciated, however, that the history of the bends contained an underlying theme that was much broader and more revealing about science and society.

From the seventeenth century, when air was first studied as matter made up of atoms, to the twentieth century, when compressed air is routinely used by engineers, divers, and doctors, the history of the use of compressed air and its effects on workers serves as a paradigm for the evolution of scientific thought. The initial enthusiasm for compressed-air technology and early lack of understanding of what caused the bends showed how discoveries relating to the Earth and its elements led to great progress as well as terrible dangers. Scientists' advances in preventing and treating the bends showed how the complex systems of the human body can be gravely influenced by the simplest of physical laws and gas chemistry. And yet these same advances show that, nevertheless, we will always be confined in our explorations by the limitations, not of the human mind, but of the human body. This book reflects my great interest in how history, science, and the pioneers of engineering and biomedical research shaped our modern world and how these lessons of the past may guide our discoveries of the future.

This book could not have been written without the great guidance and encouragement of Eric Kindwall, M.D., of the University of Wisconsin. His contribution to this book is more than that of a reviewer as he provided direction, references, and his own reflections of an illustrious career in diving medicine. I was also honored to receive the suggestions of Dr. Peter Bennett from Duke University, whose team continues to lead the world in high-tech deep sea-diving tests and physiology. I am greatly indebted to my editor, Noreen O'Connor, and to the wonderful staff of the Yale University Press, including Jean Thompson Black, Laurel Bliss, and Donna Anstey.

The research for this book required locating primary material from Europe, which I accessed via email only with the invaluable help of many individuals, including: Marie-François Prade and Odile Luguern (Inria); François Come, the Secretary General of the European Society for Engineering Education; Susan Harris of the New Bodleian Library, Oxford; Tessa Shaw of the Queen Library, Oxford; Susan McHugh of the University College, London; Sandra Cumming of the Royal Society; and the staff of the Merseyside Tourism and Conference Bureau, Liverpool.

In the United States I must thank Isabelle Millet of Yale Medical School for help with the French translations of historical texts; William Worthington of the National Museum of American History, Washington, D.C.; Carolyn Bowers of the Port Authority of New York and New Jersey; Jeff Ehmen of the NASA Marshall Space Flight Center, Alabama; Michael North and the staff at the New York Academy of Medicine Library, New York; Madeline Rogers and Norman Brouwer of the South Street Seaport Museum, New York; Tammy Gobert and Gretchen Koerpel at the Rensselaer Polytechnical Institute, Rensselaer, New York; David Stern and the staff of the Geology Library, Yale University; George Miles, Beinecke Rare Book and Manuscript Library, Yale University; Rebecca Herrera, Twentieth Century-Fox Film Corporation, Los Angeles; Wendy Phillips Kahn, Intrepid Travelers, Inc., New York; Susan Box, Medical Library of the College of Physicians and Surgeons, Columbia University, New York; Lindy Moore, Al Giddings Images, Inc., Pray, Montana; Ellen Thomasson, Missouri

Historical Society, St. Louis; and Dennis Selmont, Department of Hyperbaric Medicine, Norwalk Hospital, Connecticut.

Above all is my gratitude for my wife, Kate, for her constant encouragement, support, and interest in the realization of this project.

The Bends

# Introduction

Compressed air as used in engineering has allowed humans to span the enormous bridges, build the deepest tunnels, and design the largest harbors which catapulted the landscape and commerce of quiet eighteenth-century cities and towns into the twentieth-century metropolises we know today. Compressed air as used in diving has allowed us to leave the safety of terra firma behind and descend into a world even larger and still almost completely unknown: the sea. There, in a world with ten times the nitrogen and oxygen of the atmosphere and 99 percent of the earth's biomass of living creatures, compressed air has created a window into the miracles of the abyss for military, commercial, and leisure divers.[1] Every foot of depth was hard fought, however, as compressed air exacted its toll when divers or caisson workers attempted to resurface too quickly or go too deep, falling victim to the bends, or decompression sickness. Whether on land or in the sea, humans have harnessed the true potential of compressed air only by surmounting the enormous physiologic obstacles which cause the bends.

At the beginning of the twentieth century, these crucial properties of human physiology were understood only by a handful of visionary scientists. Today, as a result of modern laboratory methods, computer programming, and sophisticated physiologic studies, tunnel and bridge builders work in the safest conditions. Also, thousands of recreational divers enjoy the beauties of the oceanic biosphere at a hundred feet beneath the water's surface, confidently protected by years of research to protect them from the bends. Decompression accidents in construction, military, and commercial projects were, by 1996, at the lowest rate ever.[2] Divers now realistically envision staying a half-mile down in undersea communities for several months, as well as taking excursion dives off the edge of the continental shelf.[3] The future even promises the replacement of air in diving with an oxygen-enriched liquid, thus eliminating the bends and opening the sea for human exploration to depths previously unimaginable.[4] These great achievements in compressed-air use and physiology stem from a fascinating evolution of understanding and invention, debate and experiment which has intimately linked the work of physicians, politicians, engineers, physicists, soldiers, pilots, and theorists. The story of how this understanding evolved among such disparate but interconnected fields using sound observation and physical principles illustrates how new, previously unencountered diseases must be approached. The appearance of decompression sickness owing to the use of compressed air also shows that the ultimate limits of technological progress will be determined by the physiologic limits of the humans who use them.

Decompression sickness (the bends) can rightly be called the modern age's first disease. It did not exist until the machines of the Industrial Revolution, powerful enough to compress air for technical application, were adapted for human use in engineering. When it first appeared, decompression sickness was unlike any other disease encountered in workers. It had no identifiable cause, no common ailment or complaint among victims, and no reliable method of cure. It could be a nuisance, a source for joking and teasing among comrades, and a cause of permanent crippling or agonizing death. Decompression sickness was created by a new world of mechanization, a world where human beings became so intimately involved with their own mechanical creations that they both benefited from them and suffered by them.

Such relationships, as studied in the disciplines of industrial and occupational health, were virtually unheard of when the bends first appeared. Greek physicians had long ago described coal and silver miners' diseases. Physicians over the next 1,500 years occasionally warned of well-known occupational diseases and maladies. At the turn of the nineteenth century, however, there was little appreciation for the human consequences of new machinery, industrial techniques, or hazardous materials.[5] The march of industrial society left in its wake tens of thousands of human casualties. A decisive event at this time was the invention of the steam engine as modified by James Watt in the 1770s, which realized the human endeavor to harness the power of steam for practical use. This led to the invention of air-compressing engines, or the pneumatic revolution—the most important period leading to modern industrialization. James Watt's steam engine was rapidly and widely adapted throughout Europe around 1800. Air itself was being put to work, the first practical rewards of humans' long quest to comprehend and understand the earth and its atmosphere. But, compressed air was likely the first manmade hazard of the machine age to cause human suffering. A hundred years before asbestos, fifty years before radium, and thirty years before industrial dyes, there was compressed air and the bends.

No physician was prepared to deal with the pathologic repercussions of the new technology of compressed air in the 1840s. There was no medical training to understand virtually any disease of the early nineteenth century. In dealing with the bends, the physician stood as a helpless bystander to a battle waged between the mechanical and natural worlds. The disease pitted the basic laws of compressed air (mechanics, physics, and chemistry) against the profound complexity of the human body—its physiology, behavior, and biology. For nearly a century, knowledge of the human machine lagged behind knowledge of compressed-air machines, and until the two worlds were equally well understood in the twentieth century, decompression sickness continued to exact its heavy toll.

The bends occur because of two unique properties of air-breathing animals on earth: the structure of the lungs and the body's response to changes in atmospheric pressure. Our lungs and bodies evolved in an atmosphere comprised mostly of nitrogen and oxygen, in quantities which have changed little in a hundred million years. Five billion years ago, the earliest atmosphere was vastly different. More than 90 percent gaseous hydrogen, the early earth had neither oxygen nor nitrogen, yet life somehow began.[6] A billion years later, cellular life first existed, but a billion years after that there arose a cell which was the first to use the sun's energy to split a water molecule into its two component parts. One, hydrogen, was used for a multitude of crucial chemical and synthetic reactions. The other molecule, oxygen, a toxin to the cell, was expelled from it and delivered into the atmosphere as waste.[7] From this first photosynthesized molecule of gaseous oxygen began an accumulation of the waste gas which, over the next billion years, would transform the earth's biosphere. The process of photosynthesis, based principally in the algae of the ancient oceans of the globe, became the world's source of oxygen. These primordial bacteria-like cells were also able to utilize the energy of ammonia in a process called denitrification, which eventually yields gaseous nitrogen. Analyzing levels of oxidized ores and irons in ancient rocks, scientists determined that with ocean-based cellular activity, the amount of oxygen and nitrogen in the atmosphere gradually increased.[8] By 1.9 billion years ago, the partial pressure of oxygen had risen from zero to 60 millibars (mbar or 45 mm Hg); 700 million years later, it more than doubled to 145 mbar and peaked just prior to the extinction of the dinosaurs 250 million years ago at 295 mbar (221 mm Hg). Since oxygen is highly toxic at high pressures, many oxygen-producing organisms probably died off, lowering the oxygen partial pressure to today's level of about 210 mbar (160 mm Hg) about 150 million years ago.[9] Nitrogen levels gradually rose during this time but, because nitrogen was non-toxic, it tended to higher and higher partial pressures eventually becoming almost 80 percent of all atmospheric gas. Nearly all of the remaining 20 percent of the atmosphere was still oxygen, and to this day only a tiny fraction of atmospheric gas is made up carbon dioxide, carbon monoxide, ozone, and hydrogen sulfide. Together the gases combine for a total atmospheric pressure of 14.7 pounds per square inch (psi).

About 1.9 billion years ago, a honeycombed cell evolved which was not only immune to oxygen's toxicity but used oxygen as a tiny electron-carrying atom in a process known as respiration. The cell was probably engulfed by an unrelated oxygen-sensitive cell, and the two, benefiting from each other, became a single cell with an enormous survival advantage over others. These oxygen-using cells eventually evolved into modern cells and respiration as we know it today, in which carbon dioxide is produced as a waste gas. The delivery of oxygen and the excretion of carbon dioxide remain a simple task of diffusion for the organism made up of only one or a few cells. Evolving in 14.7 pounds per square inch (psi) pressure, organisms which developed an internal organization that was not directly exposed to the atmosphere survived only if a mechanism existed whereby oxygen and carbon dioxide could be actively transported into and out of the animal. From fossil evidence in Devonian and Silurian rock (about 345 to 435 million years ago), it appears that the primitive "lung in some form has existed for hundreds of millions of years."[10] All lungs of air-breathing animals, whether they exist in the water or in the air, serve the same function: to provide a surface across which the exchange of gases can take place.

The lungs of sea-level animals evolved to function optimally at no other atmospheric pressure (14.7 psi). The cells of deep-sea fish, in contrast, evolved at much greater pressures (fifty to a hundred times greater), but are still somewhat limited in habitat to the depth in which they were selected to evolve. Humans are among a small number of animals which venture into atmospheres of different pressures. Weddell seals and whales, which breathe air and reside at sea level, evolved lungs and circulatory systems which have a tremendous tolerance for changes in pressure. Seals can dive from the surface to depths of at least fifteen hundred feet for an hour or more.[11] Human lungs, in contrast, have a limited tolerance for any differences in atmospheric pressure because we evolved in the dry, flat terrain of what is now the African continent. Today's animals and plants are complex multisystem organisms, yet they depend as much on the regulated exchange of atmospheric gases as their single-celled ancestors. In few instances is this rule as strikingly illustrated as in decompression sickness. Scuba divers, mountain climbers, and high altitude pilots test the limits of our body's capacity for adaptation. These modern adventurers try but fail to resist the laws of atmospheric pressure. Our appreciation of those principles is the basis for understanding the bends and reflects a long-standing human fascination with the air that surrounds us. The vitality of air was probably first appreciated by an early hominid who gasped for breath after a hike up a steep ravine in the Olduvai plains or who realized that death always occurred when a person stopped breathing altogether. Observations by the ancients laid the foundation for our later understanding of the atmosphere and eventually led to our ability to compress air into a form that freed us from sea-level breathing altogether.

## 2

# The Discovery of the Atmosphere

Air has had a special, even mystical meaning throughout human history. The first written record of air as a *quality* was in the sixteenth century B.C., in Egyptian Ebers papyri, whose author made a distinction between "good" and "bad" air, the former being necessary to sustain life and the latter enough in and of itself to take life away.[1] The Homeric Greek thinkers viewed air as the conveyor of life as well as all things that made life bad. Inhaled and exhaled it "flew off at the moment of death."[2] Whatever air was, it intercalated itself among all aspects of life. The want of air, which afflicted few unlucky souls, was given the name *asma* and persists to this day as the word *asthma*. Such notions preceded more specific Greek words for air, like *pneuma*, which Athenian physicians used routinely three hundred years later. Air had more of a universal quality, an essential principle of life. Where the desperation for *good* air became known as *asma*, on the opposite end of the Greek spectrum was the all-encumbering nature of *bad* air or *miasma*. Miasma, to later, classical Greeks, formally arose into the concept of a perfidious evil, a gloomy sickness of treachery and vileness.[3] Such concepts of air as an infective, mysterious, and intangible force were put into sharp contrast by Greece's first material experimental thinkers in the sixth and fifth centuries B.C..

Thales of Miletus (c. 624–c. 545 B.C.), on his long walks along the fertile Nile valley, pondered what made up living things and was impressed by the vital connection between water and life. He observed that water evaporated into air, and this implied that all of life, indeed all the earth, must be a mixture of water and air and that water was the origin and destination of everything.[4] Anaximander (c. 611–c. 546 B.C.), also a Milesian, was inspired by the presence of *to apeiron* or "substance" which made up everything including the vitality of air. "As our soul, being air, holds us together, so do breath and air surround the whole universe."[5] Leucippus and Democritus (c. 460–c. 370 B.C.) later developed Anaximander's *to apeiron* theory into a simple "atomic" theory in the fifth century B.C. which held that all things were made up of countless numbers of swirling, single, indivisible units or *atoms* (meaning uncuttable in Greek). Aristotle wrongly refuted their argument, endorsing that of the Sicilian Empedocles (fl. c. 450 B.C.), who divided all substances into opposing qualities of water and fire, earth and air. Empedocles' early experiments with a water-clock established air as a tangible entity that had the ability to act on objects like any other visible matter.

Air was particularly important to Hippocratic physicians. In *Peri Physion*, his dissertation on the human need and use of air, Hippocrates stated his belief that air was a residue of undigested food and caused disease by invading the body.[6] Hippocrates pursued the qualities' relationships as espoused by Empedocles and produced the famous treatise *peri chymon* as they affected the human body, evolving later into the notion of the four humors, which guided most medieval and early Renaissance medicine. For the next thousand years, air, nature, and all properties of the body and on earth would be viewed not as the combined result of "atoms" of whirling matter but as opposing qualities and forces which both determined illness and health and defied experimentation and analysis.[7]

The absence of analytical experimentation concerning the atmosphere and the earth defined what became the Dark Ages between A.D. 1000 and 1300. With the rapid population growth of the cities throughout Europe from the eleventh to fourteenth centuries, so rose the opportunity for infection. The Great Plagues were not seen as infectious, although the plague is now known to be caused by bacteria transmitted by the common rat. (Organisms that the medieval eye could not see were thought not to exist, an attitude changed only by Anton von Leeuwenhoek's microscope of the 1650s.) Instead, air, or something in air was to blame. The geocentrism of the Middle Ages was finally shaken by the ideas of Leonardo da Vinci, Johannes Kepler, Nicolaus Copernicus, and Galileo Galilei, and the Earth was appreciated for its proper place in the universe.

After the Renaissance, Isaac Newton (1642–1727) started a revolution. By the 1600s, the Earth and the heavenly bodies were found to obey the same laws of motion, of force, and of attraction that governed the path of a falling apple. Newton's discovery of the color spectrum not only split beams of light but firmly reestablished the Greeks' atomic concept that all things were made up of components; light, fire, water, and air itself all had smaller units of matter which awaited identification. Isaac Newton escaped the "bad air" of London's Great Plague in 1666 to his bucolic home in Lincolnshire where within five years he had discovered the tenets of classical physics, calculus, mechanics, and light theory.[8] While Newton was still a student at Trinity College, Cambridge, the Aristotelian-Platonic axis of qualities and humors was being replaced by the new philosophy of quantifiable matter as espoused by René Descartes (1596–1650). The Cartesians saw the world not as intangible entities of concept but as real, interacting particles of matter. Daniel Boorstin wrote of this revolutionary moment in science: "Everything in nature . . . could be explained by the mechanical interaction of (such) particles. There was no difference, except in intricacy, between the operation of a human body, of a tree, or of a clock."[9] Newton's refraction experiments with a glass prism demonstrated that even light could be dissected and analyzed for the individual units of matter that comprise it.

Air, although invisible, was finally considered by seventeenth century investigators to be something that could be analyzed scientifically. In 1650, the Belgian Johann Baptista van Helmont (1579–1644) incinerated sixty pounds of coal into one pound of ash in a five-pound closed vessel, yet found that the vessel and its contents still weighed sixty-five pounds. Van Helmont thought he had freed from solid matter something which became, as the Greeks thought, a chaotic, indefinable, and invisible "wild spirit" that still had substance (*spiritus sylvestre*). To name his "wild spirit," he invented a word derived from the Greek for chaos: gas.[10] Evangelista Torricelli (1608–1647) believed that such gases made up the entire atmosphere and that all things on earth rested at the bottom of a huge "sea of air" which exerted its mass effect on every aspect of life. As the seventeenth century closed Robert Boyle and Robert Hooke discovered the laws governing the sea of air. When contained in a pump, air acted as a "spring," expanding to its original volume after pressure on it was released. Pressurized air, Boyle and Hooke also found, like a tightly wound spring, could produce a force over a particular distance—an entity known in Newtonian physics as "work." Such ideas were shown to be true in the 1700s by expanding a volume of water with heat into steam and using this pressure to create steam engines. These churning, belching, glorious cauldrons of harnessed power were the material demonstration that the human understanding of natural concepts, as first tested by the Greeks, could be applied for human benefit.

Thus began the pneumatic revolution: the harnessing of air and its properties for mechanical advantage. Air expansion and compression, when controlled in a reciprocating steam engine, allowed humankind to begin to master the surface and the subterranean world. It was an assumption, therefore, that compressed air would allow people to master the underwater world as well. At first, it did not. Adventurers and entrepreneurs, hoping to explore (or exploit) the treasures of the deep were the first to use compressed air at the turn of the century. These early "modern" attempts failed, just as diving attempts over the previous centuries without the benefit of compressed air also had failed. Such disappointments stemmed from the fact that, by 1810, there was still little understanding or appreciation of how atmospheric pressure and the human body were so inextricably linked. Only after the physiology of the lungs at sea-level and the laws governing the atmosphere itself could be understood, could the lungs function successfully using compressed air *below* sea level.

## Understanding the Lungs

Claudius Galen (A.D. 131–201) was one of the first to ponder the physiology of the lungs. "In all animals that inhale through the mouth from the atmosphere and exhale it again, the lung, which is an instrument of both the voice and respiration, fills the thoracic cavity. The source of its motion is the thorax. . . . When it expands, the entire lung expands to fill the space left vacant." [11] It was Leonardo da Vinci (1452–1519) who was probably the first to appreciate the need for a mechanism to bring fresh air in and to deliver waste air out. "Where flame cannot live, no animal that breathes can live." [12] Although oxygen and carbon dioxide were unknown, da Vinci and his contemporaries believed some of Galen's concepts that the heat of the heart had to be cooled and the lungs, like a chimney, served as the exhaust port. "The rapid and continuous motion of the blood producing friction on the cellular walls of the upper ventricle, as well as by its general motion. Thus, the blood is heated and subtilized so that it can penetrate the pores and give life and spirit to the members." [13]

Perhaps the greatest leap forward in this regard was the demonstration by William Harvey (1578–1657) of the circulation of blood through the heart. Like all physicians of his day, Harvey was influenced greatly by Galenic teachings and anatomic descriptions. Studying valves in the veins with Fabricius at Padua, Harvey probably first had the idea of valve-directed flow of fluids. In his lectures by 1616, he had stated that "the lungs are the noblest part of the body (with the exception of the heart) inasmuch as they are the source of the blood . . . through the lungs passes incessantly all the nutriment of the body and the whole mass of the blood; but there are some who admit nothing unless confirmed upon authority, let them learn that the truth I am contending for can be confirmed from Galen's own words, namely that not only may the blood be transmitted from the pulmonary artery into the veins, then into the left ventricle of the heart, and from thence into the arteries of the body, but that this is affected by the ceaseless pulsation of the heart and the motion of the lungs in breathing."[14]

Thus Harvey had identified the concept which illustrated the essential components of respiration. That the heart's function was not only to circulate blood to the body but through the lungs, which served as the interface between the interior of the body and the outside world where the exchange of gases could take place. Freed from waste carbon dioxide and enriched with precious oxygen, pulmonary venous blood was then pumped out of the heart to the periphery. In addition, Harvey made it clear that the heart was partitioned as well. "In this way, therefore, it may be said that the right ventricle is made for the sake of the lungs and for the transmission of the blood through them, not for their nutrition; seeing it were unreasonable to suppose that the lungs required any much more copious a supply of nutriment . . . than either the brain . . . or the eyes . . . or the flesh of the heart."[15]

It had become clear, therefore, that the lungs and the heart functioned together to provide the body with whatever was in the air and to deliver as exhaust whatever the body had produced as waste. Da Vinci was the first to document, as did Harvey later, that the orderly expansion of the muscles of the chest was associated with a corresponding increase in the volume of the lungs as they filled with air. It did not occur to da Vinci or to Harvey that it was the pressure of the atmosphere that forced air into the lungs, as the diaphragm and chest wall contracted to create a potential space to be thus filled. This was postulated by the physiologist John Mayow in the 1620s and finally demonstrated by James Carson in 1820.[16] Such discoveries illustrated that the atmosphere was a sea of gases, hovering over all of us and bearing down with a force of their combined collisional energies to create atmospheric pressure. Scientists began studying the atmosphere not just in its natural state at sea level, but in its rarified state on mountaintops and its pressurized state in water pumps. Pressurized or compressed air had long been noticed and had been used in the bellows of fourteenth-century Norwegian kilns.[17] With scientific theory fueling mechanical application, compressed air had new, revolutionary life.

Compressed air was born, therefore, after inquiries into what made up the atmosphere, how it behaved, and how it could be manipulated into performing work. Compressed air, unlike any other kind of material or agent in building prior to that time (bricks or steel, for example), had to be understood by the scientific and engineering mind before its full potential could be realized. The invention of compressed air came just in time to fuel some of the most dramatic accomplishments of the Industrial Revolution, a time of unbridled enthusiasm for mechanical progress. It was also a time of misunderstood and disregarded occupational hazards. Workers' misery throughout history has been long chronicled and long ignored by their employers. Arsenic, tin, mercury, silver, and hundreds of industrial compounds had been used for centuries by laborers who readily accepted the risks of working with such poisons in order to earn a paltry income.[18]

Compressed air was something new and different. One could not see it, nor smell it, nor appreciate any dangers inherent to it. By 1840, compressed air was used by French pioneers and created the world's first cases of decompression sickness. Furthermore, unlike any other industrial agent prior to or since, compressed air only caused disease when the worker was removed *from* it.

The appearance of decompression sickness (DCS) reflected a rapidly growing world in science and technology—usually growing at the expense of human health. With the expanded use of pressurized gas in the late nineteenth century and the ever-increasing employment of unscreened or unhealthy workers exposed to compressed air, cases of the illness increased throughout the industrial age. Only by the mid-1880s was the rudimentary pathophysiology of decompression sickness correctly demonstrated. Understanding DCS meant understanding more of human physiology, of gas mechanics and properties and of liquid physics than early nineteenth century scientists and doctors could claim. *Preventing* DCS also required a sensitivity and appreciation of occupational hazards that was rare to nonexistent in the rampant laissez-faire capitalism of the 1800s.

# The Sea of Air Around Us

At the peak of the Industrial Revolution in Europe during the late 1700s, the factory system provided steel and new materials of unprecedented applicability and strength, steam as a new source of energy, and many skilled and unskilled laborers who were desperate for work. Human architectural endeavors began to change the world more profoundly than at any time since the Roman Empire. Few other fields during the nineteenth century depended so much on human labor and transformed how society operated so completely as engineering. No single engineering agent, except perhaps explosives, helped advance modern architecture more, at the expense of human health, than compressed air.

Compressed or pressurized air was unique in that it was the first substance used by people that could injure or kill, not while one was exposed to it (regardless of how long), but only *after* one's removal from it. The invention of compressed air in the 1840s brought to fruition a desire to understand what elements or molecules made up air, how air behaved in nature and the laboratory, and how it could be used. The single greatest step in this regard was the understanding and development of the vacuum, an entity which had provoked debate back to the Athenian agora. Thirteen hundred years before John Dalton's atomic theory, Democritus postulated that empty space must exist between the atoms which composed all things.[1]

Aristotle refused to believe this or the existence of any void. "From what has been said, therefore, it is evident that the void does not exist . . . so much for the discussion of the void, and the sense in which it exists, and the sense in which it does not exist."[2] He believed that if there were a void or vacuum, light could not pass through it. The absolute rejection of the possibility of a vacuum persisted for a millennium. The so-called *horror vacui* represented a condition which "nature could not allow to exist," or so thought philosophers until the 1600s.

The Dutchman Isaac Beekman (1588–1637) was the first to write in 1614 that a vacuum *could* exist, and Galileo Galilei was told by Giovanfrancesco Sagredo in 1615 that a heated bottle which cooled contained some force which could suck water into it.[3] Galileo himself wondered why silver miners could never pump water by hand more than thirty-two feet in the air, but he died in 1642 without ever grasping the concept of atmospheric pressure. Gasparo Berti (1600–1643) had been included in such scientific discussions for years before undertaking the first experiments which led to the development of the first barometer.[4] Interested more in making a vacuum, Berti had a large spherical glass flask made with a bell inside it. The flask was attached to a forty-foot pipe. The pipe was closed at the bottom by a valve, and the tube was placed standing up in a barrel of water. The pipe was filled to the top, just below where the sphere and the bell were placed. Berti opened the valve at the pipe bottom and the water descended in the pipe, thus expanding or pulling out the air in the flask, or so Berti had hoped. A partial vacuum was created, and the bell's ring dimmed. Berti noted, however, that the column of water in the glass tube always stopped descending about eight feet below the bell or at a height of about thirty-two feet.

Galileo's student Evangelista Torricelli (1608–1647) thought it was too cumbersome to demonstrate anything easily with water, which required such tall tubes. Instead, he switched to a much more dense liquid—mercury, or quicksilver. Torricelli built a tube filled with quicksilver which he allowed to settle on a pan. He found that the column also descended to a particular level, leaving a vacuum above the meniscus similar to Berti's water column experiment. With the more dense mercury, the level was now only three feet high. Torricelli immediately grasped the significance of the observation and knew that the weight of the atmospheric air was "pressing" on the pool of mercury. This forced an equivalent weight of mercury to stay within its glass tubing. Torricelli wrote to his friend Michelangelo Ricci in 1644, "We live submerged at the bottom of a sea of air."[5] The column of water that Galileo studied in miners' pipes rose in the hand-cranked suction-pump vacuum, pressed by the sea of air, until the water-column weight equaled the "air" weight. Torricelli thus demonstrated that one "atmosphere" of pressure was equivalent to a column of water thirty-two feet high, later determined to be a pressure of 14.7 pounds per square inch gauge (psig). The relationship of air to water pressure would be vitally important to any diver, submariner, or compressed air worker two hundred years later.

Torricelli's student Blaise Pascal (1623–1662) finally demonstrated the effect of different "weights" of air (different altitudes) on a column of mercury when he and his brother-in-law placed their newly invented "barometer" on church steeples in Paris and mountains in Auvergne, France. These experiments used a column of mercury essentially no different than Torricelli's, but its level was regulated with a simple valve. Pascal found that the height of the column was about three inches lower on the Puy de Dôme than at sea level in Paris.

What interested most of these early aerophysicists was not so much the concept of air pressure but that a vacuum could be reliably created and tested. True mechanical vacuums were created with pumps first by Otto von Guericke, the mayor of the Prussian town of Magdeburg, in 1646–47. He had heard of Torricelli's work and asked how one could create a vacuum with a mechani-

Figure 1. Robert Boyle epitomized the erudite gentry which could afford to inquire into the broader questions of the human world. (Yale University, Harvey Cushing/John Hay Whitney Medical Library)

cal device. Using a cylinder and a suction pump with two flap valves he fashioned in his brother's brewery, von Guericke used his machines to create the first manmade vacuum. Although crude, the vacuum was strong enough to thwart the efforts of two teams of eight horses to wrench apart two "Magdeburg hemispheres" squeezed together by the enormous force of the atmosphere on the contraption.[6]

With experimental vacuum production possible, the single most important leap in understanding the physics of air and an organism's dependence on it came with the work of the Irish chemist Robert Boyle (1627–1691) and his assistant and friend Robert Hooke (1635–1703).

Robert Boyle was the fourteenth child of the first Earl of Cork

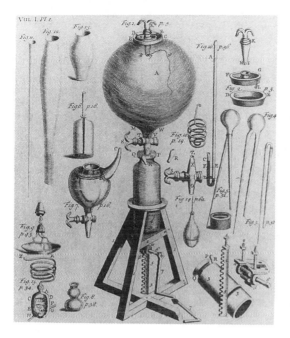

Figure 2. The air pump of Robert Boyle. In this simple flask, undoubtedly built by Robert Hooke, Boyle first demonstrated the laws of the simple vacuum and the relationship between the expansion of air and its decreasing pressure, now known as Boyle's Law. (Frontispiece of *New Experiments Physico-Mechanical touching the Spring of the Air and its Effects* . . . by Robert Boyle, 1662, Ash.C.24, Bodleian Library, University of Oxford)

(fig. 1). Touring Europe throughout his teenage years, young Robert was early enthralled by the investigative sciences and by 1645 he enrolled at Oxford. Primarily working on the chemistry of saltpeter and gunpowder, Boyle became interested in air pressure and the vacuum in the 1650s, soon after reading of von Guericke's device and Torricelli's experiments. At Christ Church College in Oxford in 1660, Boyle had his assistant Robert Hooke construct a glass bulb with a crank to lower a diaphragm at the bulb's lower end, with a series of stopcocks to regulate air flow into and out of the bulb chamber (fig. 2). Boyle was quick to notice a springlike action to the air when compressed or expanded—which led to his theories on the "Spring of Air," published in 1680 and clearly modeled on Hooke's own work of the 1670s on the metallic spring laws. In his early experiments, Boyle noted that at constant room temperature, the pressure in the air chamber rose and fell in exact inverse proportion to the volume of air he added or removed. If the air in the chamber was doubled in volume, for example, the air pressure dropped in half. This constant relationship, now known as Boyle's Law, or P (pressure) × V (volume) = constant (k), is a cornerstone of fundamental gas physics.

Modern gas physics described what gas pressure actually *was* in the "kinetic theory," described by scientists two hundred years after Boyle's death. The kinetic theory offers a molecular explanation for what Boyle and Hooke observed in their air chambers or what occurred in the lungs of a SCUBA diver under two hundred feet of sea water.

### Air and Water Pressure as Billiard Balls and Logjams

Gases and liquids are similar in that they are both fluids and both occupy the space of an enclosed container. They are different in the way their molecules interact among themselves and with the container wall. These properties distinguish air pressure from water pressure and lie at the basis of compressed air phenomena like those that Boyle observed or like those which plague divers and tunnelers in the form of decompression sickness.

Gas molecules are so small that gravity has very little influence on them. Writers often refer to the "weight of the atmosphere," as is done for simplicity in this book, but air really doesn't have "weight" at all. Rather, the pressure of the atmosphere is determined by the sum of energy transfers which occur when gas molecules collide with each other and everything else. Each molecule of nitrogen or oxygen at sea level in the atmosphere whizzes around at speeds of 1,000 mph and collides with others millions of times a microsecond.[7] These interactions can be loosely thought of as billiard balls, which independently strike other balls one at a time. Unlike billiard balls, which lose some of their energy by heating up and bending at the moment of a collision, gas molecules are *elastic* and lose no energy or shape after each collision. Instead, almost all of the energy of the molecule's momentum (mass × velocity) is transferred to another molecule which, in turn, careens off to strike the first mass it encounters. When a billiard ball hits the side of the pool table, the change in momentum generates a force over the area of contact. Gas molecules are too small to strike a substantial area of container wall alone, but combined with the millions of strikes per microsecond per square inch, the force can be substantial and *can* be measured. A handful of air at sea level contains about 400 billion billion molecules (4 followed by twenty zeros) and a measuring gauge records their combined average forces of collision as 14.7 pounds per square inch.[8] Atmospheric pressure at a particular altitude is due, therefore, to the combined forces of all the molecular collisions on a particular unit of area. At thirty miles in altitude where the air density is much less, gas molecules will collide with other molecules once every inch. At an altitude of four hundred miles, collisions occur once every forty miles.[9] With fewer collisions and fewer opportunities to change their velocity, less force per unit area or pressure exists. Robert Boyle thus demonstrated the inversely proportional relationship between pressure and volume by artificially decreasing the density of air in his pump—producing fewer elastic collisions and therefore less pressure.

Jacques A. Charles (1746–1823), a French physicist and balloonist, and J. L. Gay-Lussac (1778–1850), Parisian chemist, independently found in the 1820s that pressure or volume was proportionally related to temperature.[4] On the molecular level, raising the temperature of a gas increases the speed with which they move. Collisions occur with greater changes in velocity and impulse and therefore increased pressure is generated. Charles and Gay-Lussac's work, in its simple and concrete hypothesis, laid the groundwork for the sophisticated and abstract theories of late nineteenth-century gas physicists.

Viennese physicist Ludwig Eduard Boltzmann (1844–1906) was the first to describe how such molecular collisions and energy transfers explain the behavior of things in the real world that we see, feel, and measure. His applications of mechanics and the laws of averages to predict the behavior of atoms were at first rejected by the European scientific community. With the development of quantum mechanics in the 1900s, Boltzmann earned his rightful place in atomic theory. His mathematics, logic, and statistics, however, would have had a tough time predicting the molecular motion of the other fluid in nature: liquid.

Whereas gas behaves like a perfect system of billiard balls, liquids behave like a molecular logjam. Unlike gases, liquid molecules are very close together, so close that the jostling of one molecule influences and is influenced by the jostling of many other molecular neighbors. With closeness and compactness of molecules in a liquid, gravitational forces are significant and liquids, therefore, really do have a measurable weight. Although molecular collisions can be thought of as producing pressure, the force of a liquid's weight per unit area is much more important. Liquid molecules are so close together that a closed container of the liquid is virtually incompressible. Boyle thus found that with liquids he was not able to observe the same pressure-volume relationship he found with gases. Changes in temperature did have an effect: heating a liquid increases molecular motion and this causes some of the liquid molecules to behave more like a gas, rise to the surface, and escape the surface tension into the atmosphere, a process known as boiling.

If Hooke's genius lay in his mechanical and logical mind, Boyle's strength was his overall concept of science's connectivity to humans and nature.[11] Interested early in how air affected organisms, Boyle was the first to appreciate that removing an organism from its atmosphere of birth could have catastrophic consequences. Studying animals in a sealed glass chamber, he noted that asphyxia gradually occurred as the oxygen in the chamber atmosphere was consumed, a process usually requiring many minutes. Curiously, when Boyle removed air from the chamber and the pressure dropped to create a vacuum, the animals usually convulsed within minutes, gasping and dying. His description of the experiment is as follows: "To satisfy ourselves in some measure about the account upon which the respiration is necessary to the animals that nature hath furnished with lungs, we took . . . a lark . . . which being put in the receiver, did [several] times spring up in it a good height. The vessel being hastily, but carefully closed, the vacuum pump was diligently plied, and the bird for a while appeared lively enough. But upon greater exsuction of the air, and very soon after was taken with as violent and irregular convulsions, as are wont to be observed in poultry, when their heads are wrung off." [12]

In the *Sceptical Chymist* of 1661, Boyle speculated that the organism was deprived of a "vital quintessence." In one animal thus expired he noted bubbles of gas forming in the eyes, an eerie sign that would have tremendous significance for the twentieth-century understanding of decompression sickness. Boyle never found what his "vital quintessence" was which "serves to the refreshment and restoration of [the] vital spirit." It was, of course, not only oxygen, as discovered by Priestley more than a century later in 1774, but the ambient atmospheric pressure of the animal's surroundings upon which the animal depended for all its tissue functions. Deprived of the atmosphere of air, or the 14.7 psi of pressure in which the animal is continuously dependent, the animal suffered what can only be called the world's first fatality from decompression—a maladjustment to lower pressures than those to which the animal is accustomed. Although Boyle's air chamber "decompression" occurred after decreasing pressures from 1 atmospheric unit to *less than 1* (as also occurs in plane or balloon flights to high altitude), the same principles apply when decompressing from, for example, 10 atmospheres to 1. Such physiologic subtleties would evade scientists for another hundred and fifty years, but Boyle's work clearly established the dependence of the organism on air pressure homeostasis and the deleterious effects which can arise when the air pressure is perturbed too quickly.

Robert Hooke on his own was the first to test the effects of a rarefied atmosphere on a human being: himself. Before the Royal Society, he discussed his scientific goals and mentioned that his "air-vessel" was ready for testing. "He added that the chief design of [the] vessel was to find what change the rarefaction of the air would produce in man. . . . [The vessel] consisted of two tuns [or barrels], one included in the other; the one to hold a man, the other filled with water to cover the former . . . and contained a [gauge] to see to what degree the air was rarefied." On February 23, 1671, Hooke went in the air-vessel for fifteen minutes and reported that "a man could not endure much more than the evacuation of a fourth part of the air." On March 2, Hooke complained of profound ear pain and had found himself hard of hearing. "A candle burning with him in the vessel . . . went out long before I felt any of the inconveniences in [the] ears."[13] Although not likely developing decompression sickness, Hooke was the world's first experimental traveler to a lower atmospheric area in a laboratory chamber, which presaged the modern air chambers developed for the same purpose by Paul Bert in the 1880s. Hooke's other landmark experiments concerned respiration itself. Finding that he could keep a dog alive as long as he pumped air into and out of its lungs with a pair of bellows, he invented endotracheal intubation and artificial respiration.[14]

The attempt to understand how air pressure changes affect an animal paralleled the overall maturation of scientific experimentation and chemical analysis in the early 1700s. Equipped with an appreciation of electricity, magnetism, and pressure, scientists were prepared to grapple with nature's invisible forces. Among the most fascinating was combustion—specifically, what it was in air which allowed combustion to proceed. Without experimentation, philosophers reasoned in the 1700s that all burnable things contained a substance or *phlogiston* which actually did the burning. The phlogiston theory pervaded chemical inquiry and theory until Benjamin Franklin (1706–1790) asked his friend the Yorkshire philanthropist and politician Joseph Priestley (1733–1804) to write on the history of electricity. Priestley had been studying carbon dioxide for years, obtaining copious samples from the brewery he lived next to in Leeds. His dissolution of carbon dioxide in water was the first soda water, which led to the popular soft drink; but he was busy studying the effects of combustion on various chemicals he isolated like nitric oxide, ammonia, and hydrogen chloride. Studying how electrical discharges combusted in air, he theorized that air was made up of different substances which provided combustibility. When he heated red mercury (mercuric oxide), he found that air was expelled from it very readily and was suspicious enough to save it for further characterization. "But what surprised me more than I can well express, was, that a candle burned this air with a remarkably vigorous flame." [15]

This "air" was, of course, oxygen, and its discovery in 1774 and the ability to scientifically identify the true components of air was the beginning of a chemical revolution. Priestley himself called the gas "dephlogisticated air," because in the old theory, a gas which so readily supports combustion must be devoid of phlogiston so that the phlogiston of other things has somewhere to go. Using "dephlogistonated air," Priestly "procured a mouse, and put it into a glass vessel, containing two-ounce measures of the air from a mercurious calcinatus. Had it been common air, a full grown mouse, as this was, would have lived in it about a quarter of an hour. In this air, however, my mouse lived a full half hour." [16] The French chemist Antoine Laurent Lavoisier and his wife, Marie, coined the term "oxygen," which incorrectly reflected the belief that the gas formed acid (from the Greek, *oxys*, acid). [17]

Lavoisier was puzzled with his combustion experiments in air, as the majority of the air volume never reacted well. The unreactive component, according to the old theory, must contain a large quantity of "phlogiston" and was termed "phlogistonated air." Not a big proponent of the phlogiston theory, Lavoisier called this gas *azote* ("without life") instead. Henry Cavendish (1731–1810) later mixed "dephlogistonated air" (oxygen) with Lavoisier's azote, and it rapidly combusted into mostly a vapor of acid "niter" (nitrous oxide). The chemist Daniel Rutherford finally coined the term nitrogen for the gas, as it gave rise to niter, in 1772.

Gradually, scientists came to perceive the world as a tremendous, well-organized collection of elements. Air, by 1780, was then seen not as a receptacle for "phlogiston" but as a mixture of gases; mostly nitrogen, but also oxygen, carbon dioxide, hydrogen, and gaseous products of combustion like nitrogen dioxide, carbon monoxide, and ozone. Lavoisier believed that good air for respiration was a careful mixture of a "vital air," like the "vital quintessence" cited by Boyle, and nitrogen, with the proportions of about one to three. Using carefully constructed bell chambers, Lavoisier was able to absorb expelled carbon dioxide from test animals with chemical alkalis in the jar. This allowed Lavoisier and his colleagues to measure the first values for oxygen consumption at rest, during work, and after eating, all two hundred years before the landmark metabolic studies of the 1900s. It became clear to Lavoisier that it was oxygen which supported life and combustion. The phlogiston theory was put to rest.

John Dalton (1766–1844), father of the atomic theory, first recognized that despite their differences in quantity in the atmosphere, all the gases maintained their own contribution to the total air pressure. Dalton was interested in the elemental combination of parts to equal a given compound, the basis for the atomic theory. Air was similar to matter in that the total pressure of a given atmosphere was equal to the sum of the partial pressures of each contributing gas. Dalton's Law is as follows: the total pressure exerted by a mixture of gases is the sum of the pressures that would be exerted by each of the gases if it alone were present and occupied the total volume.[18] The total air pressure, P, was equal to the combined *partial* pressures of all the represented gases, i.e., $P_1 + P_2 + \ldots + P_n = P$. A container of air (20 percent oxygen and 80 percent nitrogen)[19] at one atmosphere (partial pressure of oxygen, 0.2 atmospheres) is pressurized to 10 atmospheres. By Dalton's Law, the oxygen still delivers 20 percent of the total pressure or 2 atmospheres. This becomes important, for example, in deep diving, where even minute traces of poisonous gases like carbon monoxide can have deadly effects at high pressures.

Several years after publishing his theories Dalton had conversations with another citizen from Manchester, William Henry (1774–1836), an analytical, skilled doctor.[20] Henry had studied heat and calories but also gas dissolution in liquids. He found that he could dissolve in water twice as much gas if the gas were condensed to *double* the pressure. In short, a solution of soda will hold five times the amount of carbon dioxide it can hold at 1 atmosphere when in a bottle pressurized to 5 atmospheres, although the volume of carbon dioxide is the same. When the bottle is opened, the dissolved carbon dioxide expands to its volume at 1 atmosphere, exceeds its dissolvability in the soda water, and rushes out of solution in bubbles. This effect, known as Henry's Law, defines how much gas can be dissolved in a fluid or tissue depending on the gas solubility in that fluid or tissue: concentration (C) = pressure (P) × gas solubility (S).

Berti, Galileo, Torricelli, Pascal, Boyle, Hooke, Priestley, the Lavoisiers, Dalton, and Henry labored away at the mysteries of the invisible gas that surrounds us, described how it behaves, and uncovered the physical laws that govern its effect on other things. The gas laws developed by these experimenters and thinkers are crucial to the safe use of compressed gas by today's navy or recreational diver or tunnel digger, because such physical principles determine almost all aspects of gas behavior and physiology in compressed air environments. The industrial world of the 1800s, however, eager to use air for work and profit, and humans for labor, paid little heed to the bell jar and test tube experiments of these eighteenth-century chemists. And laborers unfortunately learned how dependent the human body is on the simple gas laws that govern all of us. It would be a century-long lesson learned the hard way.

# The Pneumatic Revolution

The scientific endeavor to find out what made up air and how air pressure behaved in a laboratory chamber paralleled the architectural effort to use air for work. A vacuum, as Boyle used it and as Galileo noted, develops an apparent force to draw objects into it. The stronger the vacuum, the more powerful the suction force. A perfect vacuum is, as the Greeks believed, nothing but evacuated space, a container with no collisional events between molecules of gas to generate kinetic energy and therefore no capacity to perform work. The weight of the atmosphere on an empty space without an opposing pressure is the real force that provides vacuum suction. Given enough surface area, however, this force can be quite substantial. Otto van Guericke's Magdeburg hemispheres were moderate vacuums (0.1–0.5 ATA), having evacuated up to 70 percent of the air volume by hand (fig. 3). This can create an enormous amount of force when applied over a large enough surface area. The total weight exerted on the one-foot-radius spheres by the atmosphere would have been about six thousand pounds.[1] Von Guericke demonstrated this for his townsfolk by showing that not even two opposing teams of eight horses each could pull the spheres apart.[2] Vacuum suction is therefore not suction at all, but the force of air as it rushes in to fill an evacuated space or presses on the fixed walls of an evacuated container itself. Since the atmosphere usually exerts

Figure 3. The air pump of Otto von Guericke. The two men in the lower story pump out water from the container above by moving the piston (in the vat of water) up and down the evacuation tube. (Technica Curiosa, 1667, Yale Collection of Western Americana, Beinecke Rare Book and Manuscript Library)

14.7 pounds per square inch (psi) of pressure, this is the maximum force of vacuum suction that can be generated, regardless of the device used. Even modern pump evacuators cannot remove *every* molecule of air from a container. Vacuum pumps through history and today are therefore limited to the force of atmospheric pressure.[3]

In contrast to the evacuated space of a vacuum, a container of many more molecules of air than the surrounding atmosphere has many more collisional events and generates more potential energy. Stored potential energy is a source of work. More collisional events will occur with more molecules of gas (or with higher temperatures), more potential energy will be created and the ability to perform work increased. This is the basis for using pressurized air for work, for, unlike a vacuum, compressed air is limited in its work capacity only by the amount of pressure that can be generated by another force, the engine. The invention of high-pressure compressed air would require several decades of industrial advancement, essentially the development of an efficient machine to generate a high enough pressure head. The change came about when engineers thought to take advantage of water's ability to expand in volume when boiled into steam and to shrink in volume when condensed during cooling.

The expansion of steam could be used to perform work in a far more powerful way than any vacuum and, with the proper mechanics, a continuous cycle of work could be performed. Such landmark advances in the use of air and gas mechanics combined with engineering launched the modern age of machines. Thomas "Captain" Savery (1650–1715) of London is said to have observed a heated flask cool while it was turned upside down in a bowl of wine at a local tavern.[4] As the flask cooled, the red wine slowly climbed up the neck of the flask to a level higher than the level of wine in the bowl. Savery realized that as the air in the flask cooled, it shrank and the surrounding air "pressed" the wine up into the flask like a barometer. Savery had the genius, however, to assume that steam, if cooled, behaved the same way as air and created a kind of suction. Cooling huge metal vats of steam with cold water, he generated a powerful source of suction (though one still dependent on atmospheric pressure). Savery was able to thus build a suction pump which could extract more water from a mine per hour than any hand- or animal-powered device (fig. 4). His machine, though cumbersome and nonportable, had no moving

parts; it was used to quickly rescue flooded mines and shafts and was dubbed "the miner's friend."[5] It could do nothing other than pump water, despite its power. It was Savery who coined the term *horsepower* (still used today as a unit of work).[6]

Thomas Newcomen (1663–1729) of Dartmouth developed a more efficient system in which the sequence of steam expansion and condensation was connected to a lever arm and valves to create a continuous cycle of power generation, limited only by the supply of fuel to boil the water (fig. 5). The Newcomen engine could pump water more efficiently than Savery's pump, but it could not change its work with different loads, used up huge quantities of coal, and was still very inefficient—especially since a huge amount of energy was lost during the cooling phase. Still, the Newcomen engine was used for at least seventy years before any improvements were made.

Joseph Black, the Scottish physician and thermodynamicist from the University of Glasgow, was about to give a lecture on heat in 1765 using an old Newcomen engine as a model. The model was broken, however, and Black asked a young scientific instrument maker from Glasgow, James Watt, to help repair it. Watt quickly noticed that a lot of energy in making the steam was lost in the machine at each stroke as the steam heated the condensing cylinder. Watt ingeniously added a separate cylinder for steam expansion to allow it to stay hot, and this alone improved efficiency tremendously (fig. 6). By 1782, he also had the pump lever work with steam expansion as well as condensation, a sequence called reciprocation, added a "governor" to allow work to

---

Figure 4. The pump of Thomas Savery. The chamber on the left was used to produce steam, and the ones at center and right cooled the steam and produce suction. Valves controlled the flow, and to fill the space created by the decreasing volume of steam, atmospheric pressure pushed water up the tube. (National Museum of American History, Smithsonian Institution, Washington, D.C.)

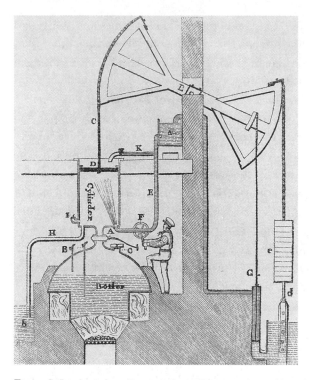

Figure 5. Improving on Savery's design, Newcomen's pump used an oscillating arm as a fulcrum by which a pumping action could be obtained. Steam entered the "cylinder" by means of a hand-cranked valve. This allowed the piston to move the large lever upward. By means of a turning crank, cool water was sprayed into the cylinder thus cooling the chamber and causing the piston to move down, pulling the lever arm with it. (National Museum of American History, Smithsonian Institution, Washington, D.C.)

be maintained with different loads on the lever and a flywheel to maintain steam production with varying loads.[7] The truly modern engine was born. First used in Cornish mines, the Watt engine fueled the Industrial Revolution and allowed a new world of machines to be built powered simply by boiling water. At 40 horsepower (about 30 kilowatts), the single cylinder engine was used in mining initially, like Savery's and Newcomen's engines, to pump

Figure 6. The first modern steam engine, Watt's engine, and its adaptation by Matthew Boulton at the Soho engine works, improved on Newcomen's by separating the condenser from the main cylinder so the cylinder could always remain hot and thus conserve energy. Watt's flywheel and governor allowed different loads to be placed. (Science and Society Picture Library)

water.[8] Later, Smeaton's pumps of the 1790s were a still greater improvement and were used successfully in allowing the construction of a harbor backwater at Ramsgate, England, by providing fresh air to diving bell laborers.

It has been long known by many children equipped with a straw that forced air could displace water from an upside-down glass submerged in a bowl of water. If the water were very deep, as in the confines of a flooded mine, pressurized air would be needed to perform the same feat. Watt's steam engine, in series with a compressor, was the first that could really be used for this purpose. Throughout the early 1800s, mine drainage was frequently accomplished with compressed air, and since a flooded mine contained no workers, no concerns were ever raised about the safety of compressed air or the health of workers exposed to it. Soon, newer nonmining applications for compressed air were developed.

For the financiers, compressed air proved to be the means by which rivers could be crossed by tremendous bridges, docks could be built, and underwater mines dug. For the laborers that worked in compressed air environments, it was the most strange and potentially lethal occupational hazard yet encountered.

## The Ocean Depths

Human efforts to access subterranean deposits of minerals are documented in antiquity, and the hazards associated with these underground labors continue today. Hoping to find additional natural resources, people shifted their attention to the sea. Water presented a particular problem for the laborer because the depth and time of work was determined solely by one's ability to hold breath. Although extraordinary feats of breath-holding were and still are reported among some South Sea Pacific sponge fishermen who routinely dive to depths of one hundred feet, breath-holding was impractical for most of the world's population. Like all other mammals, including whales, humans were forced to bring a supply of air with them into the depths.

There are reports of diving exploits by humans for thousands of years, unrecorded attempts likely for tens of thousands of years.[9] Aristotle, who was the first to describe scientifically thousands of observations of the natural and human world, recorded that Alexander the Great's army used a metal helmet which could retain air in it for submarine escapades.[10] A variant of this may have been used by now-legendary swimmers who supposedly worked on underwater anchors, cables, and snarled wrecks at the siege near Tyre in 333 B.C. Alexander the Great himself (356–323 B.C.) was said to have peered into the depths there from within a glass barrel called a "colimpha" which kept him dry (fig. 7).[11]

At Syracuse, both Julius Caesar and Pliny first mentioned a primitive device in which a diver breathed through a tube connected to the surface.[12] However, inspired air remains at atmo-

Figure 7. Alexander the Great (356–323 B.C.), king of Macedonia, conquered nations, but the subterranean world eluded him. Attempting to delve
into its mysteries, he was lowered into the sea in a glass barrel, as depicted
above by an unknown sixteenth-century Indian artist. (The Metropolitan
Museum of Art, Gift of Alexander Smith Cochran, 1913. [13.228.27]. All
rights reserved, The Metropolitan Museum of Art)

spheric pressure while the pressure on the diver's chest wall can be much higher. This is similar to the simplest of underwater breathing devices, the snorkel. This differential can create an enormous amount of breathing resistance, which can be impossible to overcome and hazardous to try. Human respiratory muscles, as studied by Mayow and Carson, have a finite ability to overcome these pressure differences, amounting to little more than 100 cm $H_2O$ (10 pascal).[13] As a result, continued attempts to inspire through these long "snorkels" can lead to airway collapse, pulmonary edema, and hemorrhage and death. Additionally, when the hydrostatic force of the surrounding water exerts itself on an unopposed thoracic cavity, blood pooling occurs, and cardiac work can increase as much as 30–60 percent. Richard Stigler in 1911 submersed himself to a depth of two meters to determine the strength of the inspiratory muscles of his chest. Without success at pulling in any air through a snorkel, despite vigorous attempts, he surfaced, desperately ill with an irregular heartbeat and heart failure.[14] He gradually recovered, but for these reasons the leather hooded devices of the Roman era by Vegetius in A.D. 375 in which divers breathed through long tubes connected to the hood, never worked.

Correspondingly, the work of respiration is influenced greatly by the turbulence of inspired air. A snorkel provides a large conduit for laminar flow. Too small a snorkel and high air flow generates obstructive, nonlaminar flow.[15] In 1680, the Italian astronomer and mathematician Giovanni Borelli was the first to develop a system to free humans from surface air sources.[16] The diver wore a huge air bag around his head with a small window for vision. Exhaled air was rerouted out of the bag to a condenser. This device never worked because with descent to depth the pressure on the bag of air and the diver's body was always the same. To inspire a given volume of air, as on the surface, the diver's chest wall would have to displace an equal volume of water. The weight of water to be displaced for a 500 cc breath is approximately twenty-five pounds for each foot of depth, and such work of respiration is intolerable below three feet for any length of time.[17] Still, Borelli's concept foreshadowed the modern self-contained breathing apparatus (scuba) developed first by Yves Le Prieur in 1925 and Jacques Cousteau and Emil Gagnan in 1943. These devices used compressed air so that the inflow of air during inspiration was provided by the pressure of the compressed air itself.

Early human adventures into the depths were thus limited by two factors: the habituation of the human body to do respiratory work at 1 atmosphere and the inability to effectively bring down a source of reusable air with each dive. For the extended periods of time required for mineral exploitation or sub-marine construction, therefore, sixteenth-century workers developed diving bells to carry them into the depths of city harbors and rivers (figs. 8, 9). This at least provided a structure within which to perform work. Roger Bacon had actually invented these devices as early as the 1240s, but in 1538 two brave Greeks descended into the depths off Toledo in an overturned metal cauldron with a burning torch. Minutes later they emerged, as audience to Spain's Emperor Charles V, with the torch still burning. Guglielmo de Lorena paid a loyal servant to carry a diving bell into Lake Nemi near Rome in 1535, the two-hundred-pound casing held up by leather straps over the diver's shoulders.[18]

By the 1600s, diving bells became more capacious, gradually able to give one man or even two enough space to perform simple tasks on the sea floor while suspended from the surface by rope or chain. The only air available for breathing, however, was the noncirculating pocket trapped at the bell dome before submersion. The nonreplenished air gradually warmed, and the ever-increasing amounts of carbon dioxide settled on the surface creating an almost suffocating working atmosphere. Most distressing was the extreme cold suffered by divers as they worked in the deep with-

Figure 8. Sixteenth-century Sturmius diving bell, first proposed by Borrelli but built and used by Johann Sturmius, the diver was submerged under the dome of a heavy iron bell. A pocket of air, decreasing in volume with each foot of depth, remained trapped in the bell dome. (From B. A. Hills, *Decompression Sickness*, copyright John Wiley and Sons, Limited. Reproduced with permission)

Figure 9. Photograph of diver under Sturmius bell. (Al Giddings Images, Inc./Charles Nicklin, 1990)

out any special accoutrements for heat conservation. Despite these severe limitations, in 1650 von Treileben was still able to sink a diving bell 40 meters in the dark waters of Stockholm harbor containing divers to retrieve valuables from the wreck of the Swedish battleship *Vasa*.[19] In 1665, George Sinclair used a three-foot-high lead bell sunk to eighty-five feet to recover the treasure of a ship of the Spanish Armada, sunk off Mullin in 1580. The American colonialist William Phipps used the same approach to become rich by salvaging a ship in water forty-six feet deep off San Domingo.[20]

The German engineer Johann Christ Sturmius observed that the trapped air in the dome of these bells obeyed Boyle's Law: the volume (V) of trapped air decreasing proportionally with the pressure (P) exerted by the surrounding height of water, or $V \times P = $ constant (k). Sturmius developed an iron-reinforced bell thirteen to fourteen feet high and first thought of resupplying divers with air in bottles. Sturmius developed a pump by 1678 and was able to force air into diving bells and exchange the compressed air of the dome. This allowed longer stays underwater but the pre-Watt hand-pumped or bellows-driven engines could not generate a sufficient pressure head to provide enough air for deeper diving bell submersions. John Lethbridge of Devonshire, England, built an enclosed bell in 1715 which protected the diver on all sides except for two small ports for the arms and a glass plate for vision (fig. 10).[21] This allowed the diver to descended to seventy-two feet without any compression of the air within. Divers found, however, that even at such depths, the weight of the water impinged directly on the seals over the unprotected arms and movement was prohibitively laborious. Edmund Halley (1656–1742)—like many of his scientific peers of the eighteenth century an expert astronomer, physicist, and chemist—was greatly interested by the secrets of the deep. Halley built perhaps the most sophisticated diving bell to date in 1691, discussing it years later as Secretary of the Royal Society in 1717. Halley's bell was supplied with air raised and lowered in lead barrels, and he vented foul air by means of valve

Figure 10. One atmosphere diving suit, 1600s. Designed by John Letheridge, this barrel allowed the diver to descend, protected from the pressure of the surrounding water, to deeper depths than diving bells could achieve. The diver's movements, as depicted here, were greatly limited and at about one hundred feet of depth, it was difficult to move the arms at all. (Courtesy, Al Giddings Images, Inc., Charles Nicklin, 1990)

controls (fig. 11). This system allowed Halley and four friends to stay at sixty feet for ninety minutes, which led him to believe that the secret to deeper dives was a more efficient pump.[22]

The British engineer John Smeaton (1724–1792) united diving with steam power. Smeaton had already been elected to the Royal

Figure 11. Seventeenth-century Halley bell. This modified bell by Edmund Halley was used by the 1690s to reach sixty feet for ninety minutes, but still limited divers to the immediate environs of where the bells touched bottom. (Courtesy, Culver Pictures)

Society at the age of twenty-nine, and after a profoundly in-fluential tour of Holland in 1754 he set about to specialize in underwater construction. Having established civil engineering as a profession in Britain in 1771, Smeaton also invented hydraulic lime, used the Newcomen steam engine to pump mines throughout Northumberland and Cornwall.[23] Smeaton recognized the problem of Halley's old bell system and quickly set about using his modified Newcomen engine to provide air to his own submarine construction. Smeaton's own bell, built in 1788, was used in Northumberland for a bridge at Hexham at which the bell was partially submerged. Smeaton's pump, a kind of force pump, was secured on top of the bell and allowed a continuous supply of fresh air. Smeaton's later design of 1791 used in the completion of the Ramsgate harbor breakwater in Kent was the first to use an iron box, or *caisson*, as a means to transport men and supplies to the work site underwater.[24] Although continuous air was supplied to the partially submerged caisson from the surface, the pump power could not generate sufficient compression of air to provide for deeper submersion of the bell. The caisson design, however, presaged modern surface-supplied diving but still provided no way for divers or materials to go freely between surface and caisson without loss of pressure.

Given Smeaton's intellect and practical grasp of the relation among architecture, water pressure, and mechanics, he was ready to make the engineering leap into modern compressed air use. His death in October 1792, however, ended his efforts, and there were few innovations in compressed air for decades. Diving bells were improved throughout the next century with improved air pumps, cooled air, and caustic soda to absorb carbon dioxide as designed by Brize-Fradin in 1808. By 1826, these changes allowed divers to work for up to ten hours. At best, diving bells allowed men to venture out from their bell, to walk or swim at depth, and then to return to the bell when cold, exhaustion, or fear required it. The problem of overcoming the great weight of water compressing the air in the work environment remained the single most significant obstacle to progress. Until both pump and caisson design were improved in the 1840s, dives deep enough for important mining and engineering projects were impossible.

An important relationship in diving medicine is that for each thirty-three feet of sea water depth, an additional atmosphere of pressure, or 14.7 psi, is exerted. This was the same relationship observed by Galileo and proved by Torricelli in the 1600s. For even modest bell dives to 66 feet (about 3 ATA), the 30 psi gauge force needed to counteract water pressure could be generated by the best Watt steam engines of the late 1830s. How human beings working in the bells tolerated such pressures would become soon tragically apparent.

# 5

## Triger's Caisson

John Smeaton's pump had enough force to exchange only the small "ceiling" of air from the top of his diving bell. For deeper submersions, Smeaton would make the following calculations: Following Boyle's Law, a six-foot chamber of air would be compressed to three feet at a depth of thirty-three feet sea water (2 atmospheres absolute of pressure or 14.7 psi gauge or psig).[1] To increase the volume of air in the bell would therefore require pressures greater than 14.7 psi, and the horsepower to sustain a full six feet of working air at 2 atmospheres of pressure for one hour was a load surpassing the most efficient machines of the time. Watt's first engines delivered thirty kilowatts and were clearly inadequate for "modern" nineteenth-century applications.[2] No steam engine through the early 1800s could accomplish even five hundred kilowatts of power. (Compare these early prototypes with the 1968 engines of the Saturn V rocket, which had a total work capacity of three hundred *million* kilowatts of power during liftoff, or enough energy to illuminate the entire city of New York for fifty seconds).[3]

The improvement in power output, using steam as the same power source, came as a result of gradual improvements in efficiency, the use of a variety of condensers, feedwater heaters, and superheaters.[4] The efficiency of a machine pump is related to energy consumed for a given amount of work output, and this depends greatly on a pump's *application* of work.[5] An inability to couple the

Figure 12. Charles-Jean Triger, French paleontologist and engineer, who was the first to apply compressed air in mining. (From G. Barral, *Le Panthéon Scientifique de la Tour Eiffel,* 1892)

Watt steam engine with other machines frustrated many entrepreneurs during the early years of the Industrial Revolution. Diving bells of the time were aerated by hand-driven or bellows-operated pumps. Each bell was a chamber with two sets of pipes for air exchange but without a formalized valve system, pressurized compartments, or locks through which men and tools could pass without losing any pressurized state that could be achieved. Even with steam power and early condensers, these limitations severely handicapped diving-bell applications for underwater work.

The first to apply the technology of steam-powered pumps with a work-chamber built to maintain the compressed air state was a French civil engineer, Charles-Jean Triger (fig. 12).[6] Triger's invention of the modern caisson revolutionized the use of compressed air in engineering. The caisson also provided the setting for the invention of an infamous new disease, known to Triger's readers of the early 1840 and 1850s as *mal de caisson.*

Charles-Jean (or Jules, according to some sources) Triger was born at Mamers in the department of La Sarthe, France, on March 11, 1801.[7] He completed his education in Paris under the well-known geologist Pierre Louis Cordier (1777–1862), chair of the Museé d'Histoire Naturelle in the early 1800s. Triger became intensely interested in all aspects of the natural sciences, paleontology, chemistry, geology, and engineering. After returning to Mamers, Triger made such a good living building quarries that by 1833 he could focus his attention on academics. He was an active paleontologist and was best known to his contemporaries for his important paleontologic surveys of strata fossils. Triger took special interest in the fossilized remains of Cambrian life found in the vast mantle of chalk uniting France and England under the Channel and rising to majestics heights at Dover. The Jurassic dinoflagellate *Amonites trigeri* was named in Triger's honor by Hebert and Eudes-Deslongchamps in 1860.[8]

Triger also studied the geologic formations of Belgium and France, where he followed up on the important work of Cordier on sources of coal in the Loire Valley. By the beginning of the 1800s, coal was one of France's most abundant resources, and it fed the belching furnaces of Paris and Nancy during the Industrial Revolution. The coal mines in the south and northeast of France were so embroiled in privatized ownership disputes, and their accessible resources were diminishing so rapidly, that alternative sites of mining were sought.[9] The Loire Valley coal beds in western France are a vast shelf of Cambrian anthracite and bituminous coal and lie close to the surface near Chalonnes. Unfortunately, most of the Loire River covers the coal beds with silt, quicksand, and mud. Even hundreds of feet away from the Loire, the coal beds were well below water level, which made access of the mineral mechanistically impossible. With the success of multiple cylinder steam engines and efficient valve systems, Triger had by the mid-1830s constructed many mines which were compartmentalized to avoid total submersion in areas of recurrent flooding. These water locks allowed the continued excavation of material when other chambers were inoperable due to mud or water accumulation.

It is not known who originally asked Triger to consider digging to the coal beds of Lyon, but his geological knowledge of the area and his familiarity with compressed air in mining probably sparked his interest in the project. Reading of a design by Sir Thomas Cochrane in 1830, Triger intended to use air under pressure to force out the water from the quicksand in the caisson. This would enable him to dig deeper, sink the caisson farther down, and eventually bypass the quicksand to reach dry earth and the beds of coal underneath. It would require a radically different mine construction, one which was completely air- and watertight, would not lose any volume of pressurized air, and would allow free access to and from the digging area without loss of compression.[10] Triger designed an iron shaft, a *caisson* (meaning box in French), which was actually a connected series of tubes reaching 1.33 meters in diameter but which could be lengthened by serially adding more "tubes" to the apex of the caisson (fig. 13).[11] The cutting edge was placed in the quicksand of an artesian well near Chalonnes and sunk to a depth of nineteen meters in the spring of 1840. The chamber roof of the caisson was sealed shut and air was pumped into the shaft to a pressure of about 20 psig. Regulated by inlet and outlet valves, a compartment midshaft acted as an air lock. In this way, the layer of quicksand was passed and a vast bed of coal was reached, overcoming an engineering difficulty previously thought insurmountable. Workers entered the air lock or chamber through the hinged door valves located at the top and bottom of the chamber. The total air pressure did not exceed 3 ATA. Work proceeded for approximately two weeks and shifts originally lasted seven to ten hours.

Triger himself entered the work area first before he would allow other workers to do so by merely climbing down the access ladder, through the air lock, and into the work chamber without any time for acclimatization. In Triger's recollection of his time in the shaft

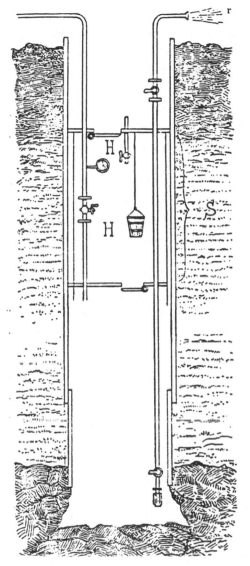

Figure 13. Triger's first caisson design, 1841. Triger's pioneering design allowed him to access a bed of coal underneath fifteen feet of quicksand in the Loire Valley, a feat that eluded all other engineers. A simple air lock allowed passage of workers without loss of pressurized air in the work space below. (From R. Heller et al., *Luftdruckerkrankungen*, Vienna: Alfred Hölder, 1900)

there is no mention of any effects of compressed air on him while in the caisson or after coming out. Triger noted in his workers, however, that "no illness was occasioned among [them] except that two . . . after seven hours of labor in the tube, experienced severe pains in the arms and knees."[12] Such symptoms did not occur immediately after leaving the shaft but "about half an hour after ascending into the open air," and many, including Triger himself, complained of a certain "breathlessness" or *suffocatio redux*.[13]

This is the first recorded observation of what Triger called *mal de caisson*, and what twenty years later Dr. Andrew Smith of the United States termed *caisson disease*.[14] Triger reported on several curious effects of working in compressed air, but his medical interpretations of the syndrome were limited. Fortunately, he and his workers finished the short project, fulfilling local contracts and allowing permanent access to the Loire coal reserves, without major injuries. Triger was an entrepreneur, an inventor, and a courageous academician who was rare among his class of employers. He refused to allow his employees access to the chamber until he had tested its safety himself.

Triger hired two physicians to supervise safety concerns on his later projects. The physicians, B. Pol and T. J. J. Watelle, made the world's first formal medical observations and treatments for a disease which no humans had previously encountered.[15]

Triger's caisson system itself was quickly applied by others in 1845 in a mine in Lourche, Douchy, in northern France where the sixty-four men employed were supervised by Pol and Watelle. Pol was a surgeon by training from the local society in Douchy but was nearing the end of his career with its associated prestige, public citations, and experience.[16] Watelle was already a member of the nearby Medical Society of Douai and was under Pol's aegis and direction.[17] The two were referred to Triger when he inquired in the region of the Département du Nord for medical help caring for any casualties at the Douchy mine, which he expected would be several feet deeper and require higher pressures of compressed air than his earlier mine in the Loire Valley at Chalonnes.

Pol and Watelle not only cared for Triger's men but represented the modern European school of medical training, which the North American doctors William Osler and Abraham Flexner would take back to the United States later in the nineteenth century after visiting German medical schools.[18] Under the old school, arising out of the guild system of the Middle Ages, physicians were trained by other physicians in disorganized, randomly guided apprenticeships. Treatments for diseases were still based heavily on Platonic-Aristotelian ideals of humoral disequilibrium, and most physicians were still reliant on versions of human anatomy first proposed in classical Greek times by Galen.[19] Barber-surgeons and medical internists began to develop their own schools by the end of the eighteenth century, but medical therapies were still highly regionalized, untested and hazardous because they were not based on accurate understandings of disease processes and their causes. With the arrival of the microscopic world of bacteriology, laboratory analysis, and the clinical exam, medical training became more formalized, university based and soundly academic.[20] Based on observational tools and empiric learning, training for physicians included data collection, assimilation, and then diagnosis before treatments were prescribed or administered. A physician surnamed François, who cared for compressed-air workers in the 1850s, expressed the sentiment well. "We stop here our conclusions from our observations on the effect of the compressed air: we are aware that this important topic is opened to broad scientific discussions; but, practitioner before all, we restrict our discussion to facts and only facts and would be satisfied if our little piece of work can be of any use to our medical colleagues at large."[21] Pol and Watelle apparently realized quickly the importance of their work for the medical community and took their observational tasks very seriously, documenting the minute details of almost all sixty-four workers employed by Triger. In their report on decompression sickness in 1854, Pol and Watelle reflected a newly established and growing discipline among physicians. In contrast to the unprincipled, empiric treatments of diseases in the past, Pol and Watelle emphasized the importance of detail, description, study, and self-teaching which is now standard medical practice. As they themselves wrote of their own work with the patients, they cited the old adage of Latin scholars, "Not simply counting but continually observing."[22]

The Lourches caisson was similar to the one at Chalonnes but somewhat larger and deeper (fig. 14). Each man had to pass a physical exam by Pol before being hired, a practice almost unheard of in industrial Europe. Triger restricted the men to fewer shifts of shorter hours, and men taken sick were discharged from the service. At a pressure of 3.5 atmospheres, and working in two shifts of four hours a day, the men seemed to tolerate the labors relatively well. The climate at the dark, muddy bottom of the caisson was humid and stifling, with little to no air venting. The exhaust from lanterns was choking and covered the lips and face with soot. Curiously, voices were thought to be higher pitched and it was impossible to whistle.[23]

Leaving the caisson was, however, a different matter. Pol and Watelle noted that upon exiting, all the men experienced a feeling of suffocation or "chokes" (the breathlessness or suffocation noted by Triger). In some, muscle pains, arthritis, and a painful itching were reported. Although Watelle seems to have escaped the ill effects of studying workers in compressed air, Pol was not so lucky. After spending several hours in the shaft in what was not his first visit, he was wracked by such severe pains of the arms, legs, and chest wall that they did not go away until the following day after a whole night of prostration, paralysis, and vomiting.[24] Gradually, Pol recovered from his illness and apparently without permanent effects.

Pol and Watelle were the first to realize that men became ill only after *leaving* the caisson and that there was some effect of the loss

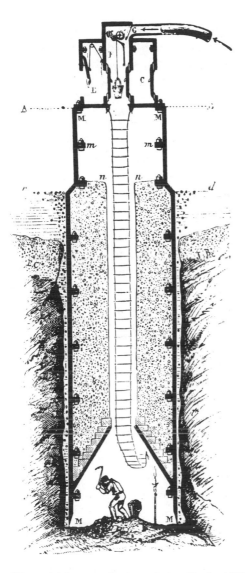

Figure 14. Improved caisson design. By the 1850s the air lock had been placed at the top, as in this caisson used in the building of a bridge in north-ern France. (From Heller et al., *Luftdruckerkrankungen*)

of compressed air on the body that was causing damage other than the extreme cooling of which men complained. They also concluded that the "the harmful effects of decompression [were] directly related to their rapidity [or decompression itself]."[25] They erroneously theorized, however, that with compression, the circulating blood was also compressed and absorbed large quantities of oxygen but that this did no harm per se. Only after *decompression*, when the high concentration of oxygen was "lost," blood became "congested" and led to the illness, and in some cases, the deaths observed. Many other contemporary physicians erroneously subscribed to some variant of the "post-caisson congestion" theory since congestion of major vessels and the spinal cord was commonly observed at autopsy. Pol and Watelle found, in the several autopsies they performed, on numerous occasions darkly congested meninges (the covering of the brain) and thickened blood in the vessels of the spinal cord and brain. Pol, despite their lack of experimental evidence, concluded that "the ill effects are in proportion to the rapidity with which the transition is made from the compressed air to the normal atmosphere." In one of the most eloquent and famous descriptions of the bends, Pol wrote, "One only pays on leaving [*On ne paie qu'en sortent*]. The *treatment* is not different from that which is usual when like symptoms arise from other causes. The first indication is to bring about a reaction, which is in turn, to be combated when it exceeds physiologic limits. . . . One case seems to indicate that the quickest and safest means of restoration is an immediate *return to the compressed air*."[26]

Despite their error in identifying the true cause of the sickness of decompression, Pol and Watelle were the first to suspect *recompression* as a means to reverse the agony of decompression. They proposed that "in order to enact a reform compatible with the needs of service, we would ask . . . to have two air locks, with one used exclusively to decompress. The compression which does not need to be as slow would be done in the other one."[27] With no scientific instruments or laboratory analysis of their own, Pol and Watelle's analysis could hardly have been any more sophisticated; if they had more time and a medical recompression chamber they may have been able even to use recompression as an actual treatment for sick men. However, after Triger's Douchy project was completed, Pol and Watelle apparently went back to their practices and several years would elapse before understanding of the disease advanced.

Experiences with compressed air and its debilitating effects upon workers returning to normal atmosphere continued in the 1850s throughout Western Europe (fig. 15). Engineers and physicians at work sites realized that the rate at which a worker exited the caissons was, in some way, related to the risk of becoming ill. Simplistic time tables for decompression began to be used, but they were difficult to enforce and gravely inadequate.[28] With more widespread compressed-air use, the spectrum of reported maladies and sickness increased and included bloody noses, ear pain and diminished hearing, excessive thirst and hunger, bloody coughs, bone pain, paralysis, intractable vomiting, bloody urine, and excruciating headaches. The ear pains described were of such magnitude on some occasions that men thought their "skulls were ripping apart," and seeing blood trickle out of the ear in several men likely confirmed these false suspicions.[29] Treatments for pains were largely topical: Cold water "ablutions," scarification cups, "anodynes" and opiate lineaments, camphorated oils, belladonna, "antiphlogistics," and Jusquiame oils all made their way into the compressed-air physicians' arsenal.[30] Still, simple recompression was not used as a treatment for the illness. Even though Pol and Watelle suspected that recompression would work, no doctor would put the theory to the test until the 1890s.

The largest work undertaken using compressed air in the 1850s was the construction of a bridge over the Rhine near Strasbourg, Alsace. François, an attending physician to several hospitals in the Haut-Rhin region and member of the medical society of Stras-

Figure 15. Caisson sunk in the building of a Stockholm bridge in 1869 has an improved work capacity and improved methods of discharging excavated material over Triger's earlier designs. (Courtesy *Scientific American*, August 7, 1869)

bourg, was assigned by the Administrative Committee of the Eastern Railway to oversee the medical care of ill workers.[31] François, like Pol and Watelle before him, observed that some workers developed various afflictions immediately after leaving the caisson, whereas in others, effects did not develop for hours. In several cases, men walked home and then fell over "as if struck by lightning," wracked by intense pains of the muscles and joints.[32] Compressed air was to be taken seriously yet some men got sick at the same pressure as men who were completely unaffected. Classifying patients into the prevailing groups of the time, François believed the "nature most favourable [less inclined to illness] is the *lymphatic;* the nature which will always be affected is the *fiery,* then comes the *nervous* and the *irritable.* . . . The people subjected to congestions, hemorrhage and those with heart or lung problems must not be exposed to the influence of compressed air."[33] Curiously, François found that patients with scrofula, a form of tuberculosis of the skin rarely seen today, were "improved by a compression at a very moderate degree."[34] François's guidelines reflected an increasing formalization of compressed-air use, a systematic control of a hazardous industrial agent in an era where such concerns were nonexistent in the industrial community.

By the mid-1860s, European efforts in bridge building and mining expanded, and compressed air was successfully used to sink caisson foundations for several major bridges on the continent. In the United States, meanwhile, fueled by an unparalleled economic expansion and immigrant influx, transcontinental and metropolitan construction boomed. In few fields was this more evident than in bridge building. Two bridge projects five years apart would dwarf anything previously attempted in Europe. The crippling aftereffects of increased compressed-air use became more prevalent, and more deaths from decompression sickness occurred between 1865 and 1880 in the United States than in all of Europe up to that time.

# James B. Eads and

# the St. Louis Bridge

As the gateway to the west, St. Louis, Missouri, vied with Chicago to be the foremost American city linking eastern commerce with western expansionism to the Pacific Ocean. No event singularly transformed this region more than the development of the railroad, and St. Louis stood to profit by it. By 1860, the city had become a major distribution and processing center, surpassed only by New Orleans and Savannah as the nation's major cotton markets.[1] Huge shipments of pressed cotton, bound for New York Harbor and the mills of Massachusetts, had to contend with one major obstacle that many St. Louisians saw as an impediment to further economic expansion via rail: the Mississippi River. Trains and heavy commerce crossed the river by ferry, but by the 1860s, high fares, transit delays, and winter hazards made the system prohibitively expensive and unworkable. A bridge over the river had been considered as early as 1839 but never materialized. Few prospective builders had any practical plans to overcome the architectural impediments of such a bridge, which in the 1840s would be the largest such project in the world.

Sheer lack of capital and the cataclysmic upheaval of the Civil War prevented any serious development of a Mississippi crossing until the St. Louis and Illinois Bridge

Company was incorporated in 1864. The bridge financiers had envisioned a golden path to the future in a metal railbed connecting Missouri with Illinois and would stop at little to see their goals accomplished. One requirement on which they would not compromise was the quality of engineer who would lead the bridge's inception, development, and creation. The German engineer John Augustus Roebling had been quietly developing steel-rope and suspension bridge technology into an art form in the 1850s with his crossings of the Allegheny River in Pittsburgh and the Niagara in upstate New York. But the St. Louis company founders had little appreciation for the technologic superiority of suspension building and thus turned to other sources.

The engineer they found would be known not so much for his sheer revolutionary genius, as was J. A. Roebling, but for his organizational and executive brilliance, his knowledge of the river, and his architectural pragmatism. James B. Eads—a dignified sea captain who had lived his entire life in St. Louis—was a true pioneer who would bridge St. Louis's past with its industrial future with three giant spans of cantilevered iron across the Mississippi (fig. 16). As the chief engineer for the St. Louis bridge project, Captain Eads had a sizable challenge. The bridge charter not only prohibited any suspension bridging but drawbridges as well, and each span had to be at least 350 feet to stretch across the 1,500 feet of the Mississippi.[2] But Eads knew the river, the rock below it, and the infamous, capricious currents that wreaked havoc on boating traffic of all types. Eads was a life-long river explorer and wreck salvager and had made more than five hundred trips to the river bottom in self-designed submarines, often to depths of sixty-five feet.[3] His geologic soundings of the river were thorough, and he had for years studied the long, gradual slope of bedrock as it tilted far underneath even the deepest mud and silt. He had never, however, built a bridge.[4]

Eads learned of Europe's most recent advances: Robert Brereton sank a pier to fifty-four pounds per square inch (psi) gauge in

Figure 16. Captain James Buchanan Eads, chief engineer of the St. Louis bridge project. (Yale Collection of Western Americana, Beinecke Rare Book and Manuscript Library)

the building of the Royal Albert Bridge at Saltash, England; John Hawkshaw built the Londonderry bridge system with a pier sunk to seventy-five feet.[5] In 1869 Eads had visited a bridge construction site at Vichy, France, and then first saw Monsieur Moreux use compressed air to sink forty piers to seventy-five feet below water level (3.3 ATA) at the Po in Placenza.[6] He discussed with Moreux's engineers the effects of compressed air on the men who labored on the project as well as the caisson and air-lock technology which was required to confine and control the compressed air produced. Neither Moreux nor Brereton could help Eads, however, in predicting the effects of compressed air required to sink what Eads calculated to be the largest caisson ever built. At 5,700 square feet and 1,250 tons, each caisson of the St. Louis bridge would have to be sunk to bedrock, at least 90–110 feet below the water surface.[7] Calvin Woodward documented well Eads's early grappling with the magnitude of the bridge. "[Eads's] own diving bell experience made him the best judge of the effects of air pressure and yet even he had no adequate idea of the peculiar results to so great a depth. He was left without any benefit from the experience of others, either in guarding against the injurious effects of the great pressure upon the workmen and engineers subjected to it, or in relieving those affected by it."[8]

Eads met and signed on two German architects, Henry Flad and Charles Pfeifer, both from the Alsace area but trained in Berlin and Stuttgart in pure mathematics and theoretical engineering.[9] Little is known of their suspension bridging expertise, but the same physics of suspension bridging applies to cantilevers, trusses, archways, or any method of bridge building: an even redistribution of changing loads and weights over the strongest segments of the structure. This mechanical law was discovered by prehistoric man: a log fallen over a stream is a kind of girder or beam bridge, supported by the banks at either end. The early Greeks found that hemicircular archways were stronger than the beam if built as an upside curve supported in the center by a keystone, because the downward forces are focused on the compressed center, whereas in the beam bridge the outward forces are directed at the structural bases. Longer archways and supports could thus be built. The girder bridge was refined in China with the development of the cantilever bridge.[10] With cantileverage, piers rise up from the base and stretch outward on both sides. (Think of the palms of both hands facing up, wrists together, holding a platter.) This structure, which focuses a longer stretch of a roadway's weight onto a narrow pier, allows the crossing of larger, commercially busy rivers where river traffic can not be disrupted by bulky or low bridgeworks. The arrangement also proves flexible for changing volume in the river. Ancient examples of cantilever bridges like at Srinigar, Kashmir, or over the Dragon River in Fukien Province, China, allow the quick currents of the flood season to pass without compromising the support structure of the bridge.[11]

Even with modern techniques of the nineteenth and twentieth centuries, cantilevered bridges are not without drawbacks. The Quebec River bridge over the St. Lawrence River, built in 1904, was intended to provide strength for the roadway as well as minimize waterway traffic jams below with two very narrow piers. During its original construction, however, the two giant cantilevers were hoisted into place, and soon thereafter the whole bridge infrastructure, twenty thousand tons of metal, collapsed and plummeted into the river with seventy-five men.[12] A second attempt with a wider bridge also fell, killing eleven. Finally, the third bridge was constructed in its present form by 1918. Cantileverage has its theoretical limits where the cantilevers fail to stabilize the lateral motions and swaying prone to extremely long sections of bridge. The maximum span for modern high-tensile steel is 2,500 feet.[13] Still, cantilever bridges reached 1,710 feet in two spans by 1890 (the Firth of Forth bridge, which is still in use today). The Commodore Barry Bridge over the Delaware River built in 1974 reached 1,644 feet.

The St. Louis bridge builders called on an earlier component of bridge design in specifying use of a truss across the Mississippi. Before cantilevers were developed during the Industrial Revolution, armed with classical geometry and the inherent strength of geometry's strongest shape, the triangle, bridge designers built a "new" kind of strong bridge: the truss bridge. The truss was first adapted to bridge design by Andrea Palladio, and Swiss designers built a wooden truss bridge 171 feet long in the 1700s. The first in North America was built over the Connecticut River in 1785 at Bellows Falls, Vermont.[14] With the development of steel in the early nineteenth century, however, the engineering possibilities increased dramatically. The triangle, strong when made of wood, was practically indestructible when made of iron and steel. Some element of the truss is used in almost all bridges, suspension or non-suspension. Steel truss bridges are favored for spans between 700 and 1,250 feet; examples include the 1,100-foot Francis Scott Key Bridge (1977) in Baltimore, the 1,232-foot Astoria Bridge (1966) over the Columbia River in Oregon, and the Oshima bridge of 975 feet between Yanai City and Oshima, Japan (1976).

The alternative to truss bridging for immense spans is the suspension bridge. Here, a roadway is essentially suspended in midair by fastening it to huge cables which disperse the weight load onto supporting towers and then to the ground. The concept had been used by ancient cultures suspending walkways from rope over the great gorges of the Yangtze and in the Andes.[15] The earliest chain-link suspension bridge was the seventy-foot span over Jacob's Creek, Greensburgh, Pennsylvania, built by the Reverend James Finley in 1801.[16] The first bridges utilizing "wrought iron" chain links to suspend the roadway were built by Thomas Telford from 1819 to 1826. His bridge over the Menai Straits in Wales, built in 1825, had a main span of 580 feet to the island of Anglesey.[17]

With suspension bridging, however, crosswind vibrations, torque, and swaying led to the failures of many bridges. The 1831 collapse of the Broughton, England, suspension bridge built by Sir Samuel Brown occurred after oscillations were started by an army of marching troops.[18] The collapse of the Tacoma Narrows Bridge at Puget Sound in 1940 stands as a modern-day example of how the forces of nature based in simple physics can rip apart even apparently invincible structures. The bridge was twenty-eight hundred feet long but only thirty-nine feet wide and was therefore highly flexible. Nicknamed "Galloping Gerty," the bridge was stiffened not by deep trusses but by two plate girders only eight feet wide. In gentle winds, the bridge swayed sideways. In a wind of 42 mph, the vibrations and torsional twisting were so severe that the whole bridge came crashing down. The loads of suspension bridging therefore require an enormous

Figure 17. Eads's or the Mississippi Bridge (1868–1880). A cantilevered truss, the bridge was the first to use compressed air to sink the two giant mid-river piers to bedrock. (C. M. Woodward, *A History of the St. Louis Bridge* [St. Louis: G. I. Jones and Company, 1881])

amount of cross-stabilization, perhaps brought to its nineteenth-century maximum by the Roeblings in the Brooklyn Bridge.

For the crossing of the Mississippi River in the 1860s, Flad, Pfeifer, and Eads decided on a cantilevered truss bridge design with three spans. Two piers would be sunk in the river's bedrock and carry the frame that would support the entire dead weight of the bridge and its traffic. Eads's calculations were based not on possible loads he had predicted for the 1870s but for the centuries to come. Its sheer stresses overestimated by tens of times, Eads's bridge was designed to be ultrastrong, stable, and safe. It would require 18,000 tons of steel (fig. 17).[19] There was no bedrock exposed at the surface in the St. Louis area, so Eads had to design a means to dig through sedimentary debris to bedrock, at least one hundred feet below the water surface. Eads adapted compressed-air technology he had learned from the French to gigantic pro-

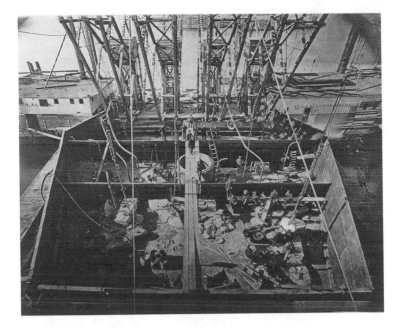

Figure 18. Surface view, Eads's caisson, showing the complicated wiring, piping, and vents needed to provide air to the workers one hundred feet below and remove the tons of sand evacuated each hour. (Yale Collection of Western Americana, Beinecke Rare Book and Manuscript Library)

portions: a wood caisson, was sunk to the river bottom and the water forced out with pressurized air (figs. 18, 19). Laborers, or "submarines," descended into the work area and began excavating sand and rock which they discharged to the surface. After several feet of mud, Eads's crews encountered hard sedimentary rock shelves, boulders, and sand, a geologic cross section of the Mississippi's past thousand years of activity. As the river bottom was dug away, masonry was built over the caisson and the weight slowly and evenly sank the whole structure until bedrock was reached and the caisson was "landed." For each thirty-four feet of descent, an additional 1 atmosphere of air pressure, or 14.7 psi, had to be pumped into the caisson to keep the work area dry.

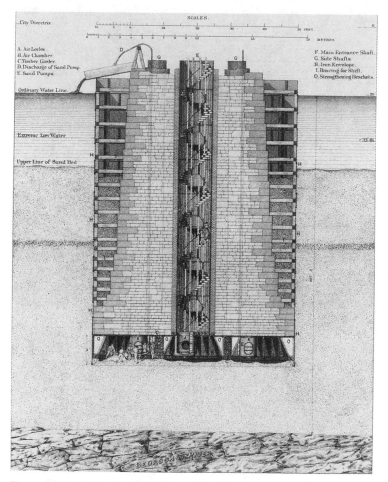

Figure 19. Eads's bridge caisson. The air lock in the early stages was at the bottom and the ascent on the winding stairs to the surface above proved debilitating to most and fatal to fourteen. (Yale Collection of Western Americana, Beinecke Rare Book and Manuscript Library)

The project began in the summer of 1868. Work proceeded around the clock, with men laboring for four- or six-hour shifts in compressed air. They felt no untoward effects until the caisson of the east pier reached a depth of sixty feet. After each work shift, the men were replaced by another team, each new worker passing through the single air lock into the caisson work area. Here, men were first subjected to the high pressures of the caisson. The lock-tender, assigned to regulate the air-lock traffic, compressions, and decompressions, would give a signal to men outside to shut the iron door to the lock. With a flip of an inlet valve, compressed air, fed by two forty-horsepower engines, rushed into the lock.[20] Some complained of an intense ear ache and others of congestion, but it was men coming out who had the real symptoms.[21] The air lock was intended to maintain the pressure of the caisson, not to ensure the safe transition of the workers to and from work area. A valve would simply be flung open and the air allowed to rush out unhindered or unchecked. After decompressing for five to ten minutes in the caisson air lock, the men climbed a winding staircase to the surface, which eventually would reach one hundred feet or, as one French visitor put it, "10 Parisian stories."[22] After work at a sixty-foot depth, where pressures reached about 25 psig, some men found that following decompression their legs would barely move. A few men experienced excruciating pains of the knee and elbow joints. Most had minor problems like itching or headache or stomach cramps.

Whenever the affliction was not that bothersome, it was not regarded as very serious. A workman walking about with a stoop and a slight paralysis was at first considered a fit object for jokes, and victims were said to have the "Grecian bend," after a walking style of fashionable ladies wearing a popular hoop skirt of the day.[23] Infamously known thereafter as the bends, the malady soon claimed several workers a day, some unable to make it up the great staircase to the surface on their own.[24]

As the caisson was submerged deeper, with pressures easily passing 25 psig, the full spectrum of decompression sickness was seen in its desperate form, remarkably similar to the symptoms seen in today's divers and caisson workers. Most bends victims complain of localized joint pain (40–92%), headache (15–33%), burning (20%), itching (15%), and a host of nonspecific problems: earaches, wheals in the skin, emphysema or trapped air in the skin, numbness, deafness, shortness of breath and the "chokes," nausea, vertigo, and bleeding from the throat and ear.[25] Severe cases often progress to involve the nervous system and can cause walking difficulties, paralysis, stroke, convulsions, and death.[26] These ailments continue today as the frightful hallmarks of decompression sickness type I (non-neurologic or minor cases: DCS I) and type II (neurologic or major: DCS II).[27] Today, with advanced medical support, deaths are rare. For the St. Louis bridge company and its workers, however, each day brought more patients who had a disease no doctor had seen before. They were clearly harmed by compressed air but in no obvious or straightforward way. Without any training or information regarding this new disease, the doctors could provide no effective medical care.

At first, the bridge company looked into the most popular contemporary treatments for what was thought to be primarily an arthritic process. They provided the men with bracelets and anklets of galvanic armor made of zinc and silver which were advertised and used in those days for ailments ranging from rheumatism to dropsy.[28] Many workers were led to believe that the armor provided protection from the compressed air, but in fact it did little except to maintain the superstitions of the men and probably the bank account of the local manufacturer. A certain "Abolition Oil" gained popularity as well for the intense itching and knee pain often reported, and one of the foremen claimed it "worked like a charm."[29] As the caisson sank deeper, the number of men with the bends increased. Some used a Volta/battery to deliver harmless electric shocks, fruitlessly attempting to relieve partial limb paralysis.

After nearly nine months of work, the east caisson hit bedrock on February 28, 1869, at 93.5 feet below the surface with an air pressure of forty-four pounds per square inch, or almost 4 atmospheres (absolute).[30] Thereafter began the laborious work of filling in the chamber with concrete. On March 19, 1869, a young man, who seemed to be healthy and was apparently feeling well after decompression, climbed the great staircase to the surface, where, after gasping for air and staggering a few moments, he collapsed and died. This was Eads's first fatality. Horrifically, six men eventually died within ten days, all given the diagnosis of apoplexy (stroke).

Desperate to prevent further tragedy as well as preserve the morale of his men, Captain Eads asked his personal physician, Alphonse Jaminet, to supervise and establish medical regulations (fig. 20). Jaminet began working on March 31 but had already been a frequent visitor to the caisson and recorded many effects of compressed air on himself. He had developed a curiosity about how liquids boiled and how boiling points were changed by air pressure.[31] He therefore had a personal interest in the workings of compressed air and created an on-site infirmary for men to rest and for those stricken with the bends. Built on a barge, the floating hospital provided berths, mattresses, and blankets. The men were required to lunch at the abutment, and hot beef-tea was provided to all. No one was allowed to leave the work area until one hour after the work shift was over. The men did not easily heed their doctor's advice, and Jaminet often voiced his frustration seeing men "hurry ashore . . . to bar rooms and places unfit for any man employed in such exhausting work."[32] Such disregard by workers for the rules regarding decompression continued throughout the nineteenth century and into the twentieth.[33] Charles T. Thackrah, the founder of occupational medicine, described the working men and their desire for alcohol not as a means to improve their lot but "to drown the ever-recurring idea, that they are, from their occupation, doomed to premature disease."[34]

Figure 20. Alphonse Jaminet (painting by A. G. Powers, 1870) had elevators for Eads's workers installed and caisson air locks placed at the surface. Jaminet's infirmary was the first of its kind in the world to provide on-site care for workers injured on the job. (Courtesy, Missouri Historical Society)

Jaminet observed that upon entering the air chamber, his pulse and those of other workers raced for several minutes in the heavy air before normalizing. He was most impressed, however, with the numbing coldness he experienced as he underwent decompression.[35] During one episode, Jaminet was decompressed after a stay at 45 psig for two and a half hours. The chief engineer who was regulating decompression merely locked the work chamber door and swung the valve wide open, allowing a drop from pressures as high as 45 psig as fast as the inch-diameter outflow pipes would accommodate. Jaminet was shocked by the ear-splitting howl of the air as it escaped out of the air lock and he knew enough to ask the lock-tender to slow the flow of air. Still, Jaminet (and the lock-tender) completed the decompression in three and a half minutes total.[36] For Jaminet's two and a half hours in compressed air at pressures over one hundred feet of sea water, his decompression was about two and half *hours* less than that recommended by modern diving tables.[37] Jaminet complained of severe cold and stomach pain almost immediately, could barely climb the staircase, and struggled home, where he became paralyzed for hours and unable to walk for days. Jaminet thereafter believed that the rates of *compression* as well as decompression adversely affected the men, and he began requiring minimum time in the chamber for both, decreased the shift lengths, and established rigorous physical examinations before and during each worker's shift.

Several men who talked themselves out of the physical exams later died after only hours in the caisson. One man died suddenly after working at the same pressure for three months. One twenty-year-old had himself smuggled into the caisson by his friend and became ill the first day. Ordered by Jaminet to stay home, he returned to be smuggled in again two days later. On the second occasion he was brought up from the work chamber unconscious on his friend's shoulders to be taken to the hospital where he gradually recovered. Changes were eventually made after Jaminet and Eads conferred: an elevator was installed, and it greatly relieved the burden of climbing the long, winding stairs.[39] The air locks were made larger, and the decompression chamber itself was brought to the top of the elevator shaft so the men could decompress at the surface after climbing the ladder.

Another mysterious observation was that some men worked for weeks, day in and day out, without any symptoms whatsoever and then suddenly had an attack. Exposed to horrible decompression schedules, these workers' bodies became so acclimated to pressurized air that even the brief decompression called for in Jaminet's tables often did not harm them. Workers who suffered the bends on the first day or two at a given pressure tolerated deeper depths and higher pressures without any symptoms. After about ten days' shifts in the caisson, Jaminet's men developed a tolerance for quicker decompressions. Assignment elsewhere on the bridge for as little as five days would make the same men lose their tolerance. The habituation of tunnel or caisson laborers to compressed-air work still has not been completely explained.[40]

Jaminet's next goal, after establishing the regimen to care for workers once stricken with the bends, was to prevent the bends altogether. As Pol, Watelle, and François had done thirty years before, Jaminet tried to regulate the transit into and out of the air lock. Where these French investigators regulated the transit time *out* of the caisson, Jaminet thought it was *compression* that was to blame and first regulated the time *in* to the caisson. "The most important [factor] of all . . . is the duration of time to remain in the air-lock when going *into* the caisson which should always be at the rate of one minute for every three pounds of pressure to the square inch. . . . When coming out, . . . the duration of time should be at the rate of one minute for every six pounds/in$^2$."[41] This type of uniform decompression is continued by some today, although with much slower, scientifically determined, rates.

The pressure at the base of the east pier would for a time reach nearly 50 psig.[42] According to Jaminet's schedules, the men underwent decompression at that pressure for twenty minutes after a two-hour shift.[43] (For men working a two-hour shift at 60 psig, in about 135 feet of sea water, the U.S. Navy would require a total ascent time of four hours, or twelve times *longer* than what Jaminet's tables prescribed.)[44]

It is amazing that more workers did not fall victim to the bends. Many men became habituated to multiple decompressions and somehow escaped the bends entirely. By the completion of the east and west piers in March 1871, nearly 25 percent, or 91 of 352 men, suffered some ailment after decompression. Thirty were hospitalized, at least thirteen died, and two were apparently crippled for life.[45] Jaminet performed several autopsies at the City Hospital on those who died either immediately after decompression or after a stay in the hospital. These post mortems were not nearly as complete as modern examinations, and on one occasion the only finding reported by Jaminet was that the patient had "had evidently no dinner and only a light breakfast."[46]

Many physicians in the St. Louis area were fascinated by the goings-on underneath the surface of the Mississippi as the bridge went up and more and more men were stricken with the strange malady of the bends. Some of these were amateur theorists who claimed to understand the pathophysiology of the bends. Dr. E. A Clark, a City Hospital associate, believed the bends resulted from the forcing of blood into vital organs by the action of compressed air on the body surface resulting in the bloody congestion observed in the brain, liver, and intestines commonly seen at post mortem.[47] Others thought it was largely the effect of the carbolic acid expelled into the work air by the candles illuminating the caisson.[48]

Jaminet not only studied the victims of the bends and performed several autopsies of fatal cases, he was attacked by the bends himself. Jaminet's theories were not well-based on sound physiologic principles but focused on preventing what he saw as the cause: heat and "vital energy" loss. He reasoned that under 4 atmospheres of pressure (5 ATA), as was present in the east pier caisson, the body's *metabolism* is obligatorily increased as well.[49] This led, he believed, to the sweating, hyperventilation, increased urine production, and exhaustion he observed in the men in the work area (these effects were more likely related to the poor convective cooling down in the caisson). "All the liquids of the body are surrounded by a certain amount of their own vapor and gases exist in cavities throughout the body. . . . In coming from an pressure of four atmospheres, there [is] four times the normal amount of such gases. In passing through the air-lock, three-fourths of this [gas] . . . escapes by expansion . . . and [carries] heat from the interior of every organ. Where the gases [are] more abundant, the loss of heat would be the greatest, viz., in the abdomen."[50]

Jaminet's observations of cerebral and spinal cord of blood congestion at autopsy also stemmed, he believed, from a similar heat loss-related phenomenon. "By coming out of the air-lock . . . the circulation of the superficial parts of the body are [changed and causes] the temporary increased arterial pressure of the blood on the brain and spinal cord . . . If [one] increases [this pressure] . . . on the brain . . . by bringing on a too sudden reaction by the use of stimulants, hot baths, or electricity, it [results] in . . . fatal cases."[51] This reasoning may have led Jaminet to prohibit the use of common "stimulants" at the site like batteries, hot baths, galvanic armor, cocaine, or alcohol. Instead, the doctor encouraged the men to wear warm jackets and get enough post-decompression rest. "Every person . . . ought to wear a flannel undershirt and a flannel belt wrapped at least twice around the abdomen. . . . [One should] leave off all his superfluous clothing after entering the air-chambers . . . but to put [them] on again, [and] add more, such as an overcoat or blanket . . . after coming out of the air-lock."[52]

Jaminet based his writings on no scientific studies, and he wrongly believed compression was equally to blame in establishing the higher vital energy as was decompression in "depleting" it.[53] His decompression tables for various depths, however insufficient, were a starting point, and his on-site hospital for those injured on the job was the first company-owned hospital for industrial workers who sustained occupationally related injuries. The "hospital" was in service at least fifteen years before local cotton mill owners in Lowell, Massachusetts, built what is generally considered the world's first company-owned hospital in the 1870s.[54]

# The Roeblings and

# the Brooklyn Bridge

While the piers and abutments of the St. Louis or "Eads Bridge" were nearing completion in 1869, work began on the even larger Brooklyn Bridge. By the mid-1800s, New York and Brooklyn had emerged as the largest and busiest cities of eastern seaboard commerce. Between 1850 and 1870, New York commercial shipping had grown so quickly that it nearly surpassed Liverpool, England, as the world's busiest port.[1] Transportation between New York and Brooklyn was, as in St. Louis, still by ferry. The infamously strong East River currents and its muddy bottom precluded any serious bridge crossings until John Augustus Roebling submitted plans to the New York–Brooklyn Bridge Company in the early 1860s. The story of the Roeblings and the building of the Brooklyn Bridge is one of personal tragedy and triumph. Although superseded by the behemoths of today, the Roebling bridge still stands as a testimony to the imaginative spirit, technological genius, and resourcefulness of its creators.

Born in Muhlhausen, Germany, J. A. Roebling (1806–1869) trained in Berlin at the Royal Polytechnic Institute and left for America in 1831 with his brother, Wilhelm.[2] Originally settling outside Pittsburgh, the Roeblings were engaged in farming and established an agrarian town which, settled by artisan Germans, was to be a new utopia called Saxonburg. Roebling's convictions about his promise in the New Land were notoriously strong.[3] It is no coincidence, therefore, that with his belief in the American Dream and pride in the history of freedom which established the country, Roebling named his first son Washington.

Somewhat dismayed by the drudgery of agrarian life in rural Pennsylvania, and the allure of practically anywhere else, J. A. Roebling eventually took a job as an engineer and surveyor for the Pennsylvania Railroad, specifically, the Portage Railroad System. In this endeavor, canal barges were hauled over the Allegheny mountains in a series of inclines. This work, reminiscent of the fictional legend of Werner Herzog's *Fitzcarraldo*, required heavy rope hawsers to do the dragging. Although the system was efficient, the hawsers or rope cables were expensive and had a limited life span. Chain link was clearly stronger but was thousands of times heavier and thus impractical. Roebling recalled an earlier German design of rope made from strands of iron wire, and his proposal to furnish the portage system with wire rope was immediately accepted and put into practice with great success. Roebling's product was so profitable that he went into full-time production initially at his Saxonburg complex.

Roebling's first encounter with a suspension bridge was the structure crossing the Bavarian river Regnitz in Bamber, which he had seen while still in Germany.[4] His design and successful implementation of a suspended canalway over the Allegheny in Pittsburgh inaugurated Roebling as America's premier builder of suspension bridges. His later triumphs were roadways over the Allegheny River in Pittsburgh in 1858, across the terrifying, two-hundred-foot high gorges of the Niagara in 1855, and over the Ohio River in Cincinnati in 1866.[5] All of these bridges utilized the potentially limitless strength of suspension technology with the necessary stabilizing safeguards that would avoid the disasters that befell English and French suspension bridges during the 1850s.

His design for Brooklyn was theoretically the same as previous accomplishments except increased to unheard of proportions. Cross-stabilization, counter-torsional suspension, and stiffening of the roadway would be preserved with the use of tens of thousands of cables fastened to each other, to the main cables, and to the roadway into a web of architectural beauty and stability. The sheer magnitude of the project, however, if it did not faze Roebling, disturbed many in the bridge company. These concerns he confidently assuaged with the great accuracy of his calculations, his tremendously detailed plans, and his personal bombast and Prussian bravado for which he was known.

Instead of the mid-river piers of Eads's truss bridge in St. Louis, Roebling proposed a single, enormous stretch of roadway sixteen hundred feet across to be suspended from two towers to span the East River. The height of the towers, the length of the single span, the depth of water, the size of the caissons, and the estimated dead load capacity of the roadway were all of a magnitude never before attempted anywhere in the world. The bridge still ranks in the top ten in the world for its class. The entire roadway was, and still is, suspended by four enormous cables of bound wire rope, four feet in circumference and 3,500 feet long, wound from nineteen smaller cables of wire in parallel which in turn were wound from 256 still smaller wires in parallel. For four cables, each 3578½ feet long and made up of 5,434 wires, a total of 21,736 miles of wire was used.[6] To support these enormous cables, two romanesque towers 276 feet high would be built on the New York and Brooklyn sides. The entire weight of the bridge and its traffic would be transferred to the towers' bases and on whatever material the towers rested. This ideally would be bedrock and core sampling by Roebling revealed that the bedrock under the East River lay at least eighty feet below sea level and was far harder stuff than Eads had ever encountered at the muddy bottom of the Mississippi. The use of compressed air would, of course, be necessary and, like all things with Roebling's bridge, the caissons used to sink the towers were larger than any other created to that time.

J. A. Roebling's dream of seeing his bridge completed was tragically never fulfilled. One day in June 1869, Roebling was studying the construction of his east tower from a vantage point on a pier when a barge collided with the jetty. Caught off balance, Roebling slipped and his foot was caught in the pier and crushed. As was common in the days before antibiotics, the foot wound became gangrenous. Competent but overwhelmed surgeons amputated the foot near the toes to stave off further spread of the infection; in true Prussian manner, Roebling refused to be given an anesthetic. Nevertheless, the gangrene spread, and he developed full-blown lockjaw and then tetanus. Within a month, Roebling eventually succumbed in an agonizing death with a horrifying type of seizure known to the medical world as opisthotonos.[7] In these death throes, the body is drawn up out of bed, the feet extending back to touch the shoulder blades, and the face stuck in the sardonic grimace of spasm just before consciousness is lost and respiration ends.

Washington Roebling (1837–1926), having begun the project with his father, assumed the role of chief engineer without delay (fig. 21). Trained in the United States at the Rensselaer Polytechnic Institute in Troy, New York, he had served in the Union Army from 1861 to 1865. Serving at the Battle of Manassas (Second Bull Run), Antietam, Chancellorsville, the Wilderness, and Spotsylvania, he was also a crucial member of the team that defended Little Round Top at Gettysburg from John Hood's Texas militia.[8] He designed Fort Sedgwick, one of the infamous labyrinthine trenches at Petersburg, Virginia, used by Grant's Union Army in 1864.[9] After a war record of almost unparalleled perseverance and luck he left the military, in one piece, as a twenty-seven-year-old lieutenant colonel (some would refer to Roebling ever after as "Colonel"). Roebling rejoined his father's engineering projects and this almost immediately included the designs for the Brooklyn Bridge.

Figure 21. Washington Augustus Roebling, son of John Augustus Roebling, accomplished his father's dream to be the first to bridge the East River. (Courtesy, Photograph Collection, AC 18, Archives and Special Collections, Rensselaer Polytechnic Institute)

Roebling was as equally gifted in engineering as his father. Young Washington completed the bridge project over the next fourteen years while carrying the burden of J. A. Roebling's reputation on his shoulders. His devotion, to his father, his memory, and the bridge would nearly cost him his own life.

The trademark task of the chief engineer is to organize the technological expertise and architectural prowess and ingenuity that are essential to any project, but especially one of such pioneering scale as the Brooklyn Bridge. Roebling knew well the importance of running a team much like an organism; a central, directing force relying on the precise, well-executed decisions and commands of the periphery. The caissons, like all foundations, were, for all intents and purposes, the entire bridge. Without a successful sinking of these two structures, the entire bridge would be a colossal failure. For such a task, he chose three engineers whom he knew very well. Francis Collingwood was a close friend of Roebling's at the Rensselaer; William Paine, self-taught in surveying and draftsmanship, served in the Union Army engineers and was hired by the Roeblings soon thereafter; and Wilhelm Hildebrand was a gifted German engineer and master designer.[10] The Brooklyn caisson, like the St. Louis design, was a wooden box with a "roof" of ten solid feet of North Carolina yellow pine, calculated to eventually hold five tons per square foot, but, on one occasion, proved itself to be five times that strong (figs. 22, 23).[11] The caisson was three times as large as Eads's (*three hundred* times as large as Triger's) and it covered more than sixteen thousand square feet or 1.4 acres.[12] The compressed air to fill these giants was to be supplied from compressors ten times as powerful as those used by Triger in 1839. Built by the Burleigh Rock and Drill Company of Massachusetts, the steam engines had a combined power of 120 horsepower and were able to keep the caissons pressurized almost continuously for a year at pressures up to 65 psig.[13] The cutting edges of the caisson were protected by a metal sheath and wedges were used to prevent slippage once

Figure 22. The launching of the first East River caisson in Brooklyn. (Courtesy, *Scientific American*, July 9, 1870)

Figure 23. A model of the Brooklyn caisson. (Courtesy, National Museum of American History, Smithsonian Institution, Washington, D.C.)

a

b

c

Figure 24. Down in the caissons of the Brooklyn Bridge: (a) wedges are cut in the dim candlelight to support the caisson cutting edge as rocks are pulled away; (b) workers excavate rocks and mud, while one leaves the chamber through the ceiling to be decompressed at the surface; (c) decompression times early in the project were solely at the whim of the air lock operator. (*Scientific American*, November 12, 1870)

digging was underway. Like the St. Louis bridge, with each foot of mud and dirt removed, more masonry was piled on top and the wedges were kicked aside allowing the caisson to descend deeper toward bedrock. Unlike Eads's bridge progress, Roebling's first caisson was exceedingly slow—six inches per week [14]—due to the presence of tremendous boulders and rock. These had been brought down from Canada within glaciers in the last Ice Age. Each had to be dynamited, chiseled, or pulled out of its prehistoric resting sites for any descent to be possible. The air lock was, as in Eads's bridge, placed at the bottom of the air shaft, and carbolic acid fumes from gas lamps were a continual hazard (fig. 24).

The digging began May 21, 1870 and after almost a year of

digging on the Brooklyn side, the New York caisson (west side) was towed into place and sunk. Workers in Brooklyn had heard of the plight of the Eads bridge laborers but were somewhat mollified in knowing that the Brooklyn caisson would not have to go as deep as the one in St. Louis. Still, by the time the Brooklyn caisson hit bedrock in February 1871, with a pressure at 37 psig, many men had been struck by some of the more mild symptoms of the bends: muscle aches, headaches, dizziness (the staggers), limb and joint pains (the bends), shortness of breath (the chokes), and itching (the niggles). Very few had severe neurological damage. The New York caisson went far deeper and was much more dangerous.

The New York caisson was very similar to its Brooklyn counterpart except it was eight hundred square feet larger and equipped with some improvements to facilitate the digging. The Manhattan side of the East River, which is really a long tidal inlet connecting the Long Island Sound with Upper New York Bay, has a bottom much different than its rocky Brooklyn counterpart. Here, Roebling would have to dig much deeper, through many more feet of hard sand before bedrock was reached, about one hundred feet down, farther than even Eads had gone. The digging went almost *twenty times* faster than that long first month in Brooklyn.[15] However, at the depth where the Brooklyn caisson hit bedrock, at forty-five feet (about 2.3 ATA), men in the Manhattan caisson started getting sick.[16] Still going down, the pressure increasing, and bedrock deeper still, Roebling grew concerned that there was a real chance for the bends to get out of hand.

In April, 1871, the Roebling's doctor, Andrew H. Smith, was asked to supervise medical care at the bridge site (fig. 25). Smith had trained in New York in general and throat surgery and after serving in the Union Army became an attending surgeon on staff at both St. Luke's and Manhattan Eye and Ear Hospital.[17] He was elected a member of the National Academy of Sciences as well as the prestigious Society of Health in Berlin, Germany. He

Figure 25. Andrew H. Smith was the official physician to the New York and Brooklyn Bridge Company. Setting minimal times for decompression, he greatly improved the safety of the workers. (Courtesy, New York Academy of Medicine Library)

represented the state of New Jersey in the autopsy team which in 1881 extracted the bullet shot into President James Garfield's back by Charles Giteau. Andrew Smith, like Eads's physician, Alphonse Jaminet, before him, established strict standards for worker health and built an on-site infirmary and bath facility.

Smith found many of the same side effects as Jaminet did. With more extensive university-based training in animal physiology and natural sciences, however, Smith's observations and explanations for what illnesses and symptoms he recorded were substantially more scientific. Jaminet ascribed the hyperventilation and fast heartrate of workers to an increased metabolism driven by compressed air. Smith, however, correctly thought the tachypnea secondary to higher ambient carbon dioxide pressure, a consequence of inefficient gas lamp fuel combustion in the caisson.[18] The fast heart rate observed in workers initially entering the chamber, Smith believed, was caused by the increased peripheral resistance produced by the high pressure. Pulse tracings, too, Smith found, decreased in amplitude owing to the same effects. Both Jaminet and Smith noted that after twenty to thirty minutes, the heart rate slowed, sometimes to rates slower than baseline. It was Smith, though, who correctly assumed the effect due to a gradual increase in the arterial oxygen pressure as the worker respired the compressed gases.

Smith had other observations in common with Jaminet but correctly explained the cause of increased body heat (decreased convective cooling in the humid caisson heat), and excess "perspiration" (water condensation). Smith explained the polyuria as due to increased renal clearance of urea secondary to decreased skin excretory functions in the compressed, humid air.[19] The polyuria, known today as compressed air diuresis, is due to the decreased secretion of antidiuretic hormone (ADH) by the brain, an effect also produced by alcohol.[20] ADH acts on the water-absorbing portion of the kidney and its absence leads to the production of copious, dilute urine.[21]

Smith could identify no logical cause for the increased appetite seen in most workers in compressed air (first described by C.-J. Triger and later Jack Hughes in 1852 at the Rochester, England, bridge site). Whereas Jaminet's notion was an increased whole body metabolism, Smith wrote, "I can scarcely believe such an increased interstitial change [occurs] without giving rise to a marked elevation of temperature and other symptoms denoting unusual chemical activity. . . . If the [metabolism] of tissue is really greatly accelerated through the influence of compressed air, it should be apparent in a more prompt healing of wounds."[22]

Belief in the benefit of compressed air for a host of diseases had begun as soon as air pressure was discovered in the 1600s. Smith's wound-healing "experiments" involved sets of pigeons allowed to heal wing lacerations during a six-day stay in the caisson. Control birds resided in the on-site hospital, dutifully watching over their human counterparts. After six days, Smith carefully inspected the bird's wounds for differences in infection (discharge or redness), closure, and scar. Studying only these simple and inaccurate measures of wound appearance, he found no gross differences.[23] This fueled Smith's skepticism of Jaminet's notions of increased tissue energy and other contemporary fads extolling the benefits of compressed air in many different ailments. This belief continued without scientific basis through the early 1900s and was resuscitated by modern medicine in the 1980s, principally in the form of hyperbaric oxygen therapy (see chapter 14).

Washington Roebling supervised the building of the great caissons on a daily basis, and, like Triger, would often go down into the caissons to test the working environment before the day's digging could proceed. In 1871, while excavation was nearing completion, Roebling suffered what was probably his worst case of the bends, and he was unable to "shake off" the agonizing pains of the legs, chest and arms seen in type II decompression sickness.[24] Tormented, he lay in his Brooklyn Heights home, unable to move for days. The accident left him an invalid in a wheelchair and emotionally exhausted. Unable to travel to the site itself, Roebling was confined to his apartment, overlooking the site with a telescope, frantically sending messages to the engineers via his wife, Emily.

While the Brooklyn caisson hit bedrock at forty-five feet below sea level in late December 1871, the New York side was still digging. Soon, forty-five feet had been passed and eventually, as the excavated sand gave way to the immense weight of masonry on top of the caisson, the seventy-foot mark was reached. The excavated material, though not bedrock, grew more difficult to remove, slowing progress and increasing the risks to the workmen from the bends. At seventy-one feet, men started dying. Smith was essentially powerless to stop this, though he persevered in giving the men comfort by any means at his disposal.[25] Among the workers, however, there was a pervasive general suspicion, perhaps rightly, that nothing within the physicians' or engineers' power could help.

Smith's study of the victims of decompression sickness was thorough, and in a speech to the College of Physicians and Surgeons in New York in 1873 he suggested that the malady be called caisson disease. He described a vast array of symptoms, organized into severity, from dermatitis and pruritus to various paralyses and death, all in a methodical, pathophysiologic manner later included as a chapter in a nationally distributed textbook on internal medicine. His autopsy findings revealed *congestion*, or engorgement of the brain and spinal cord, as "constant lesions," similar to Jaminet's observations but applying standards approaching modern techniques.[26]

Despite Smith's clinical and basic scientific acumen, he fell victim, like Jaminet, to supporting his own theory as to the cause of caisson disease without well-controlled experimental evidence. The body, he reasoned, was a compartmentalized structure with different inherent tissue densities but in "perfect communication throughout . . . and filled with a mobile fluid which is free to change its locality in obedience to any force which acts upon it. . . . When the surface of the body is subjected to an even pressure . . . the fluid blood retreats from the surface to the centre and accumulates there until an equillibrium is reached. . . . Firm and compact structures will be congested at the expense of those more compressible. . . . Structures within closed bony cavities are congested at the expense of all others."[27] With these three tenets, Smith reasoned that upon decompression, vessels congested with blood will "not readily empty themselves of the excess blood which they contain. Especially will this be the case in the brain and spinal cord where the conditions are most favourable for the production of congestion." [28]

The correct observation that disease severity was proportional to time in compressed air led incorrectly to his theory that "congested vessels will lose their contractility in proportion to the time their muscular fibres have been upon the stretch" and, thus, "less readily . . . resume its normal condition when the pressure is removed." [29]

Like Jaminet before him, Smith's greatest asset was his patient care. Although therapies for the bends included morphine, ergots, atrophine, tourniquets, hot water, cold water, electricity, alcohol, leeches, douches, and even strychnine, he did establish safer decompression schedules. Smith furthermore designed but did not build a chamber specifically for the purpose of recompression in which a bends victim would be placed on a stretcher and reintroduced into a compressed air environment. While the patient was nourished and bathed, the pressure would be, he proposed, slowly lessened over hours until 1 atmosphere was reached. Smith unassumingly had picked up on Pol and Watelle's ideas of 1853 and designed the world's first medical recompression chamber.[30] It preceded by several years a design of Ernest Moir's, which was used in the Hudson River Tunnel project between 1889 and 1890. Moir's chamber reduced mortality more than twelvefold and established therapeutic recompression as a mainstay for all future compressed-air projects; its physiologic principles remain essentially unchanged to this day. It may have been true that Smith thought of recompression as a form of therapy as an afterthought since his report was made to Columbia University years after his work at the bridge was completed. Some workers had suspected that recompression could provide therapy during the project, but were loath to attempt it when they felt so ill. Those that *did* reenter the caisson were thought of as heroes.[31] Unlike any other kind of occupational injury in which the victim removes himself from the harmful area, as suggested by Ellenbog four hundred years earlier, these nineteenth-century workers pursued relief by going back *into* the caisson, much to their happiness and onlookers' amazement. Had Smith used a medical chamber for recompression he would have given scientific credence to such measures and could have prevented the deaths that did occur.[32]

With the New York caisson still sinking, Roebling grew concerned. While no deaths had ever occurred in St. Louis while Eads was at seventy feet, Roebling already had three. He struggled with the reality that continuing to bedrock, perhaps another thirty feet, would result in even *more* deaths than in Eads's whole project. With the rate of descent now slowed 75 percent to a foot per week, it was not known how much longer it would take to dig through the natural concrete of the sandy bottom.[33] Roebling did not wait to find out.

From the microscopic appearance of the sand, which suggested that it had not been geologically changed for a million years, Roebling carefully reasoned that it was not likely to change in the practical future. In a move of unparalleled confidence in geologic processes and architectural principles, Roebling ordered the digging of the New York tower to stop. It had reached seventy-eight feet or 34.7 psig (3.4 ATA). The New York tower rests on sand, as it has for more than a hundred years, where it will likely rest for a thousand more.

With the caissons down and the towers up, the building of the superstructure commenced, and over the next ten years thirty-two miles of cable would grace the top of the towers and the roadway, all attached to the shores by steel and cement anchors each weighing sixty thousand tons.[34] The Brooklyn bridge was a milestone in architectural development and stands as a testament to the intimate relationship of form and beauty. But in terms of bridge-building, it was but a stepping stone. For the even larger bridge and tunnel projects of the next eighty years, the bends would have to be conquered first. To understand the bends—what caused it and how it could be prevented—would require the painstaking work of researchers and thinkers who were among the first truly modern laboratory and experimental scientists.

Ever since Triger employed physicians to care for injured or "bent" workers, users of compressed air exhibited a growing concern for occupational safety and workers' health. Smith and Jaminet cared for the largest population of men exposed to the injurious effects of compressed air at that time and went farther than anyone in establishing criteria, however insufficient, for safety standards regarding its use. Modern-day standards became possible only twenty years later when the scientific causes of decompression sickness were characterized. The prevailing attitudes toward worker safety in bridge building was far ahead that of any other field in engineering, labor, or industry. The on-site hospitals run by Jaminet and Smith were the first of their kind in the industrial world and reflected the pioneering attitudes of Eads and Roebling in minimizing injury and sickness on the job, one hundred years before the Occupational Safety and Health Acts (OSHA) in 1971.

Smith and Jaminet represented a transition in American medicine. Their medical training was solidly based on the European schools of science: of physics and thermodynamics, chemistry and the atomic theory, mathematics, and natural history.[35] Darwin's *Origin of Species* was published ten years before the Brooklyn Bridge was begun in 1869, and by the bridge's completion in 1883, Robert Koch had isolated the tuberculosis bacillus. Although they are among the first "company physicians," Smith and Jaminet provided no lasting medical or scientific breakthroughs regarding decompression sickness or any other malady. In the strictest sense, they were amateur researchers, pursuing scientific questions out of their own interests but fueled by sometimes misguided principles which would not have withstood debate or quantitative laboratory methods. Neither published their observations in the literature of the day and neither had had laboratory training in how to correlate their clinical observations or work. Without the national or international dissemination of clinical observations, the audience for any impressions Smith and Jaminet had would be limited indeed.

By the end of the nineteenth century, an unprecedented period of architectural activity occurred throughout the western world.[36] No place in the world was more impressive in this regard than New York City. Attempting to accommodate a torrential influx of immigrants and fortune seekers, the city was also caught in the strangling grip of commercial traffic that threatened to derail its growth. Those with foresight knew that one place could still be exploited to increase human transportation to and from the city: the underground. With compressed air, the earth under rivers and bays became newly accessible and allowed the construction of tunnels which stand to this day as a record of human achievement.

∞

# Tunneling Underground

# and Underwater

With the East River successfully bridged, Brooklyn would
never again be the peaceful bucolic farmland its Dutch
settlers once knew. It grew explosively into America's tenth
largest city in its own right by the 1930s. The East River
soon required a second span, the Williamsburg Bridge,
two miles north of the Roeblings' creation. (Built three
years faster and for less money, it has needed more repairs
than the great East River bridge ever has.) With Brooklyn
conquered, Manhattan now needed pathways to increase
commerce with America to the west. The only two ways to
get commerce and people onto the island were by either
sail or ferry. With the only train access to Long Island over
the Brooklyn Bridge, Manhattan still lay entirely cut off
from the rest of the United States by rail. The bodies of water that brought New York to its commercially preeminent
position also threatened to isolate it, as the country relied
more on overland transportation. The East River may
have been deep and a mile across, but the Hudson River,
to its west, was even deeper and twice as wide. The Brooklyn Bridge spanned sixteen hundred feet over water but
required the same distance for its New York approach
alone. Manhattanites feared that their island would be
overcrowded with viaducts and thruways. Ernest William
Moir (1862–1933), a tunnel engineer trained in En-

gland, appreciated the uneasy, continued struggle between a city's growing need for commerce and its space in which that commerce would take place. On an island like Manhattan, space was limited. The subterranean world, however, held promise. As Moir once explained, a tunnel "does not obstruct like an opening bridge, and permits easier gradients than a high level bridge [and] is to be preferred where impervious cover can be found or where the maximum depth from the surface of the water to the bottom does not exceed 85 feet, or a water pressure equivalent thereto."[1]

Tunneling through rock that is above sea level is technically challenging, but tunneling beneath a river, sea, or ocean is, even by today's engineering standards, miraculous. Submarine tunneling is, however, an art as old as commerce itself, thought to have been practiced on the grand scale by Babylonian kings to connect their temples with their royal palaces, which were separated by the Euphrates River.[2] Ancient techniques of tunneling remained unchanged for four thousand years until the advent of steam power, explosives, and steel in the Industrial Revolution. Techniques for tunneling underwater continued to improve throughout the nineteenth and twentieth centuries, and today reality approaches what was once science fiction. With remote control, engineers can now guide robotic drills through rock in compressed air tunnels many feet deeper than human workers can tolerate.[3] The advances in technology with engineering and compressed air paralleled that of the nineteenth-century bridge builders as did casualties from decompression sickness. It was in tunneling, however, that the first treatment for decompression sickness—recompression in a medical lock—was achieved. It was in tunneling, as well, that compressed-air work was first nationalized and systematically regulated. Compressed air introduced the government in its growing role of securing protection, maintenance, and improvement of worker health in almost all occupations.

To reach Manhattan by train was no easy feat. The New York

and Harlem Railroad and the New Haven Lines handled all the rail traffic from the island north to Yonkers and New England and Boston. Going east meant a trip over the East River or Williamsburg Bridge. Going west meant crossing the Hudson by ferry to the train depots on the New Jersey shore in Jersey City, Weehawken, or Hoboken. A bridge west over the mighty Hudson would need to be twice as long as the Brooklyn Bridge, cost nearly five times the $13 million the Roeblings spent, and require access to the island somewhere downtown, necessitating the razing of at least four square miles for the viaducts alone.[4]

A tunnel was early on a desirable alternative, and a wealthy mine and railway builder from San Francisco saw under the Hudson River another gold mine. Dewitt Clinton Haskin (1833–1900) was not an engineer; in fact, he had never built a tunnel let alone anything underwater. Haskin was simply an entrepreneur who looked beyond logistics, details, and even reality. Haskin, along with his associate Trevor W. Park, established the Hudson River Tunnel Company in 1873 with $1–million of their own capital, confident that money would pour in from the rest of the world to complete the $10–million project.[5] Although Haskin knew little if anything about tunnels, he was excited by Roebling's use of compressed air in the early 1870s. Haskin reasoned that instead of a solitary caisson digging down through the river floor, the caisson would be the base from which Haskin would dig a tunnel underneath the river floor, clear to the other side of the Hudson, into downtown Manhattan. In order to conquer the world's most formidable bay, Haskin should have taken advantage of the most modern, advanced tunneling techniques. His failure to do so eventually spelled financial disaster and cost honest men their lives. While Haskin dreamed of millions, more methodic designers, who would eventually be the conquerors of the Hudson, were perfecting the craft of tunneling which to this day remains almost unchanged since the 1880s. Unfortunately, decompression sickness would leave countless men in the wake of progress as rivers were crossed, cities linked, and the railroad continued its unquenchable race outward.

Marc Isambard (King) Brunel (1769–1849), the builder of England's Royal Albert and Clifton bridges and also of the ship *Great Eastern*, was also a tunneler.[6] Possessed of a penchant for marine zoology, Brunel had both despised as well as marveled at the sea worms which threatened to destroy every pier in London and every wooden ship on the seas. The teredo worm, the culprit in many a ship's death, burrowed through wood with its strong head and mouth, leaving in its wake a tunnel lined with exoskeletal debris and dried lubricant.[7] The beautiful efficiency of nature, thought Brunel, could be applied to the urban world of steel. Using a rectangular, iron head or "shield," a teredo-like tunnel could be built in which the shield would be marched forward leaving behind shells of cast-iron cemented in place by masonry. Brunel patented the shield in 1818 and first used it in 1825 to attempt completion of London's Thames Tunnel, abandoned half-finished by Vazie and Trevethick in 1808.[8] Brunel's concept was sound, but the water above the tunnel forced its way into the tiniest crack or gap of the ironworks. Imminent collapse threatened every moment. One day in January 1828 six men died as the Thames rushed horrifically into the tunnel, and the project ground to a watery halt.

Naval officer Sir Thomas Cochrane read of the Thames Tunnel disaster in the *Times* and reflected upon his experience with canal locks and diving bells in marine salvage. Discussing the subject with his scientific friends, Cochrane came up with the idea of using compressed air to fight the tremendous head of water above the tunnel. Cochrane approached Brunel, who was intrigued by the idea, as he had been when he was first introduced to compressed air in concept by his friend Dr. Colladon in 1828. Brunel recorded in his diary of March 7, 1831, that he "went to the Patent Office [and] found that Lord Cochrane had taken out a patent for some improvements in the Art of Mining—which is the scheme he has for tunneling—this plan is to force air in the excavation under a diving bell, in proportion to the head of water or the weight of ground that must be supported."[9] Had Brunel actually used the idea, he would have been the first man to do so, as well as the first to complete a tunnel under the Thames, ten years *before* Triger used compressed air to mine for coal in the Loire. Unfortunately for the workers, Brunel's unprotected shield advanced through the mud of the Thames without compressed air at all, sustaining four more floods until Brunel finally reached the other side in 1842. With all of Brunel's work and the losses of life claimed by the tunnel, it is tragic that Brunel never designed *access* to the tunnel for trains. Until the East London Railway completed the job in 1865, sixteen years after Brunel's death, his tunnel was used only by pedestrian traffic. However, Brunel's "shield" technique was born, and engineers all over the world used it.

James Henry Greathead (1844–1896) and Peter W. Barlow completed the famous Tower subway of London in 1869 using a modified Brunel shield. Unlike Brunel's rectangular shield, Greathead's was round and fit like an enormous thimble over the metal finger that was the tunnel shell.[10] Advanced into the London mud by screw jacks, the shield used compressed air to maintain a constant force against the water-head. The shield and compressed air were united by Greathead in his building of the City of London and Southwark Subway between 1886 and 1890. Sir John Fowler, who designed the great Firth of Forth bridge in Scotland, helped with this venture, which stretched three miles under the Thames and was the first of London's Underground rail system. The threat of water pressure throughout the project was constant but relatively small, as only 14 to 20 psig of compressed air were required to maintain a dry shaft. As a result, Greathead's men suffered almost no ill consequences of the bends.[11]

So much air escaped out of the roof of the cutting edge of the tunnel through the gravel and silt of the Thames bottom that huge compressors were used to send thousands of cubic feet of fresh air into the tunnel every minute; this kept the work atmosphere relatively free of soot and smoke. Greathead's assistant, Ernest W. Moir, thought that the relationship between clean air and their workers' immunity from the bends was causal remarking that "as the pressure increases purity must be greater . . . [and is] the great necessity. The impurity never affects a man while below but only after he comes out." [12] It is difficult to understand why Moir was convinced that atmospheric pollutants caused the bends only after one was removed from them, but his appreciation for good working environments and the conditions which caused the bends would benefit the rest of the engineering world. For the Southwark subway workers, luck was clearly on their side. Progress was so fast, in fact, that the *Engineer* of June 7, 1889, claimed that "the system of construction made this [tunnel] an easy task" and catapulted Greathead to worldwide fame. [13] Greathead's use of electric rail impressed the world, and even the engineers of New York and Pennsylvania began to take notice. Everyone, that is, except DeWitt Clinton Haskin.

Haskin had begun digging the Hudson River tunnel in 1879 with a thirty-eight-foot shaft sunk near Jersey City to allow tunneling west, back to 15th Street in Weehawken, and east out under the Hudson. The tunnel was made up of two elliptical tubes, about eighteen feet high and sixteen feet across. As they tunneled Haskin's men encountered soft silt, with the water of the Hudson only thirty feet above them. Compressed air at 18 psig was used initially, generated by huge steam engines and compressors in the New Jersey side shaft, and good progress was made.

Since the height of water at the top of a tunnel is less than the bottom, the air pressure must always be high enough for the lower edge of the tunnel. If air escapes from the upper edge, the pressure being greater than the waterhead, this can result in "blow-outs" of tremendous force. During the construction of the East River Tunnel in March 1905, a blow-out occurred which led to a leak.

One of the foremen, Richard Creegan, began stuffing bales of hay into the holes in an attempt to stem the flood, as was the common practice of the time. (As engineers Hewett and Johannesson explained, "Held by his fellows from being swept away, [the tunnel man] will stand by the blow, cramming sacks, coats, shirts, mud, hay, sawdust or anything into the hole, hoping that it may become choked. If forced finally to retreat by the rising flood all that can be done is for the men to get back along the runway to the emergency lock and to lock themselves out.")[14] Suddenly, a tremendous gust of compressed air came into the tunnel and sent Creegan flying out the hole into the mud with only his feet visible to his colleagues. His fellow workers, terrified, thought he was doomed until a second gust of air erupted and ejected Creegan through the five feet of mud upward into the river itself, through fifteen feet of water and to the surface atop a geyser of spray. Creegan was hauled aboard a fishing boat by stunned sailors and lived to tell the tale.[15]

For Haskin's men on a warm July day in 1880, a similar loss of pressure back in Weehawken near the shaft sent the entire tunnel pressure plummeting. At that time, the tunnel pressure was only 14.2 psig, only one-tenth greater than what was needed to keep the thirty feet of water out. The Hudson emptied into the tunnel, first a trickle, then, as noticed by assistant engineer Peter Woodland at the tunnel head, a terrifying crashing torrent, sweeping the men toward the single air lock to the work chamber. Eight men got out and nineteen remained. What Woodland and his men's thoughts were in those last moments will never be known, but Woodland instinctively shut the air-lock door on himself and his coworkers. Perhaps he realized that neither he or his nineteen men would survive, and he hoped to save the other eight men already in the lock. Or perhaps the weight of the rushing wall of water may have slammed the door on them, trapping Woodland and the nineteen others, who all were drowned. Their lives are a testament to the heroism of those leading such pioneer work or a reminder of the hazards which needlessly and recklessly took human lives too often. Eight months later, after the tunnel was pumped dry and the pressure reestablished, work resumed in January 1881. At 22 psig, Haskin finally relied on other techniques, much to the satisfaction of his chief engineer, General William Sooy Smith. Using the "pilot tube" technique developed by his Swedish engineer Anderson, the workers advanced a mini-shield beyond the working tunnel to excavate under the safety of the tunnel roof.[16]

As the tunnel crossed under the deepest part of the river, the pressure head increased to 30 psig. Able to control the waterhead, to remove even the biggest boulders, and to quell the anger of his supporters, Haskin was, however, unable to control nitrogen. (Paul Bert had identified nitrogen as the causative gas in decompression sickness as the early 1870s in his Parisian laboratory. Three thousand miles away, this single essential observation would not be known to the American engineering community for another twenty years.) In the entire building of the St. Louis bridge, thirteen men died from the bends. In Brooklyn, just three had. Not even three-quarters of the way to New York, the Hudson River tunnel claimed a death rate from decompression sickness of one man a month or 25 percent of Haskin's workers per year.[17] The underpaid workers, with no physician and no medical facilities, wondered if this tunnel was worth it after all. Eventually, so did Haskin's supporters. By 1887, the project seemed dead.

Having just conquered the Thames, the British had become world experts and had been asked by New York financiers to consult on Haskin's tunnel. Sir John Fowler and Sir Benjamin Baker recommended finally that the shield technique of Greathead be utilized. The English firm of S. Pearson and Son were selected as the contractors and arranged the transportation of a Greathead shield from Scotland across the Atlantic to New Jersey. W. R. Hutton represented the designers and slowly the eighty-ton behemoth was pieced together more than two thousand feet away from the Weehawken shore. Ernest Moir represented the Pearson firm

Figure 26. Ernest William Moir (1862–1933), inventor of the first medical air lock. (*Literary Digest*, vol. 36, 1908, 573)

and immediately set out to improve the efficiency of the digging (fig. 26). At 35 psig, the tremendous grinding and excavation toward New York began again.

Moir was aware of the one-in-four annual mortality rate among Haskin's workers due to decompression sickness alone, and he was surprised to note that "nobody had seemed to care anything about it." [18] It is not known if Moir had heard of Andrew Smith's medical lock idea after completion of the Brooklyn Bridge or had read of Pol and Watelle's recompression theories from the 1850s. Nevertheless, Moir instinctively felt that recompression was the solution for these unfortunate men. As the working chamber was now more than a half mile from shore, the medical lock, which was essentially a

Figure 27. Moir's first air lock, 1889–90. Although imposing, the medical lock saved the lives of countless workers with severe bends in the Hudson River tunnel before it was applied throughout the industrial world. (*Literary Digest*, vol. 36, 1908, 573)

boiler on its side, was placed back on the opening to the Jersey shore (fig. 27). A man overcome by paralysis or pain was placed in the chamber and repressurized to two-thirds of the working pressure. The pressure was then lowered by one pound per minute "or even less, the time allowed for equalisation being from 25 to 30 minutes." "Even in severe cases," reported Moir, "the men went away quite cured."[19] Although these were the world's first theoretically correct and effective treatments for decompression sickness, Moir's decompression times from the lock were still very risky. For a worker at 35 psig having worked an eight-hour shift, a thirty- to forty-five-minute decompression in a lock at 23 psig was twenty-five times too fast by present-day standards.

Moir himself had referred to his "homeopathically" based treatments: the belief that disease can be cured by exposing a patient to a dilute form of a substance that ordinarily can produce similar symptoms.[20] Of course, it was not homeopathy at all, since it was not compressed air per se that was causing the bends but rather decompressing *from* it.[21] Moir's understanding of decompression sickness was very similar to the nonscientific theories proposed by Alphonse Jaminet, that of compressed air putting the body into a kind of metabolic overdrive. "It appears to me," Moir wrote to London's Society of Arts in 1896 (echoing Leonardo da Vinci's theory of pulmonary physiology), that

> when a man goes into compressed air he is in the same condition as a furnace under forced draught. Suddenly three or four times the weight of oxygen is passed over his lung surface, or through its furnace, the system gradually assimilates itself to that increased oxygen and more combustion goes on. When he comes out, however, there is a sudden reduction in the amount of oxygen, the forced draught is shut off, as it were, and, as it is in the case of a furnace, there is a production of carbonic oxide, or an accumulation of carbon through an insufficiency of oxygen to burn it up. I think there is much the same effect in the case of compressed air. The carbon goes on accumulating in the blood, and after the man has come out he is actually poisoned by the effect of carbonic acid or carbonic oxide.[22]

By the time his report was published in 1896 the effects of decompression sickness were clearly related to the liberation of nitrogen, not oxygen hypermetabolism. Yet Moir appreciated the ill effects of air contaminants, first appreciated by Triger in 1839 who saw the soot of candles burning in compressed air as a constant nuisance. These poisonous products of combustion are a continual hazard, even in modern tunneling or caisson work. The Holland tunnel in New York completed in 1937, circulates ten million cubic feet of air per day through huge air purification vents a train could drive through.[23] The compressors for the Clyde tunnels in Scotland could provide free air at twenty-nine thousand cubic feet per *minute*.[24]

Moir thought that air impurities, like carbonic acid, were particularly dangerous at high pressure. Although not the cause of the bends, carbon monoxide and other products of combustion have a markedly enhanced toxicity at high pressures. Moir viewed the caisson worker as a kind of living soda bottle, since "the blood under the increased air pressure actually absorbs carbonic acid, as does the water in the manufacture of aerated [soda] waters which may bubble off when the man comes out and stop the circulation. . . . I have seen a man's veins opened whose blood was so thick and black that it had to be squeezed out. . . . This man did not recover and was one of the cases which I think would have been cured by the lock if we had had it from the first."[25]

Having subdued the bends, Moir and his coworkers had less luck with subduing the Hudson. Huge boulders were soon encountered with nothing above them but silt and the river itself. The tunnelers found that blasting, therefore, was impossible, and the great shield from Scotland stopped in the mud sixteen hundred feet from New York. By 1891, Haskin finally gave up.

But tunnel fever had gripped New York. The Pennsylvania Railroad Company was busy at work digging its own tunnels under the Hudson and the East rivers to make Manhattan the great hub of long-distance railroad travel, centralized by the great Pennsylvania Station.[26] One of the railroad's chief engineers, Charles Jacobs, would eventually use compressed air and modern shield techniques to rescue the challenging Ravenswood gas tunnel from Manhattan to Long Island in 1894. There Jacobs had seen compressed air pressures as high as 55 psig, an almost unbelievable record of lethality(which now would be illegal).[27] The colorful Georgia entrepreneur William G. McAdoo approached Jacobs in the 1890s to see if he could make any use of the abandoned Haskin tunnel. Under the aegis of the newly formed Hudson and Manhattan Railway Company, Jacobs got the old Pearson/Greathead shield moving again by 1902. Using super-hot kerosene torches to literally burn the silty overhead in the tunnel into hard clay,

Figure 28. Hudson River tunnel construction, 1889. The cutting edge of the shield finally employed in 1889 to complete the first tunnel under the Hudson River by 1905, establishing what is now the PATH tunnel system. (*Scientific American*, July 5, 1890)

Jacobs was able to stabilize the working head and remove any sized boulder below that stood in his way.[28] Eventually, Jacobs was able to blindly shove the shield forward, pushing aside mud and clay at a rate thirty-five times faster than even Pearson and Moir (fig. 28). The Haskin tunnels begun back in the 1890s from New York were eventually reached by the long tubes coming from New Jersey by 1905 and the Hudson was finally tunneled. The two tubes eventually reached Sixth Avenue and 33d Street and downtown to Morton Street in Manhattan and eventually carried commuters on the Port Authority Trans-Hudson or PATH system, which today carries 50,000 people daily.[29]

The Pennsylvania Railroad Company tunnel system eventually evolved into the main subterranean access to New York City by train for the Washington, D.C., to Boston northeast corridor. The Haskin-Jacobs tunnel may have reached New York first, but the

Pennsylvania tunnels were much larger. Where the Haskin Hudson River tunnel maximum air pressure reached 35 psig, the Pennsylvania tunnels reached 37 psig under the Hudson and 42 psig under the East River. Correspondingly, between 1904 and 1909, 3,692 tunnel workers required treatment for decompression sickness,[30] and the care became systematic. In addition to the locks, rooms were built for the diggers or "sandhogs" near the air locks, with heated lockers to keep clothes dry, copious supplies of hot coffee, and couches to rest on. Moir's medical lock itself was divided into a recompression chamber and a conduit through which a doctor could come and go. Moir's great contribution to the health of diggers' lives did not go unnoticed, especially by the sandhogs themselves. Years later at a reunion, former tunnelers of the Hudson project presented Moir with a model of the air lock Moir had built, complete with working valves and doors.[31] Moir's lock saved the lives of many men and dropped the monthly mortality rate from 25 percent to 1.66 percent. Still, a 1.66 percent yearly mortality did not take into account those men crippled or injured by the bends who survived but were unable to work thereafter. To lower mortality to zero and minimize morbidity, engineers and physicians over the next twenty years developed strict decompression tables, with compliance required by law. Working with the diving community, the great navies of the United Kingdom and the physicians and engineers around the world, real advances in worker safety, it was thought, would surely follow. Embroiled in union contracts, disputes, and the ever-present profit motive, advances in diving safety and aeronautics, naval and air force decompression tables would leave the tables that had been used and promulgated by the caisson and tunnelling world in the dust for sixty years.

Across the ocean, in France and then England, where the Industrial Revolution was first ignited, now began the modern scientific revolution. Where American achievements like the light bulb and the telephone answered questions of necessity, European accomplishments were stronger in addressing questions of reason and theory. Laboratory analysis and new scientific techniques would reveal the particle laws of light, electromagnetism, microbial life, and human physiology. No longer would any element of nature be viewed as an indecipherable quality to be accepted as is, but as a dissectable, yielding entity. The reasoning and inquiry of Descartes and Harvey were thus truly modernized and the quantitative nature and the approach to natural questions dogmatized. The science of the late nineteenth century was ready to tackle any problem its scientists could identify and put to controlled, measurable experimentation. Decompression sickness, first brought on by machine just fifty years before, would slowly be unraveled by this new generation of researchers and thinkers. Then, the greatest achievements in treating the disease would surely be within reach.

9

# Paul Bert and the Cause

# of Decompression Sickness

When French physiologist Paul Bert (1833–1886) was completing his classic experiments on the effects of barometric pressure in the 1870s, he recalled a certain practice of Mediterranean fishermen off the southern coast of France (fig. 29). When catching fish in especially deep waters, the fishermen were known to use a wooden or metal spike to pierce the swimming bladders. This apparently prevented rupture of the bladder and the other visceral organs and delayed spoilage when the fish were brought to the surface.[1] Bert recognized that fish, like humans, were confined to a particular ambient pressure, and it was the expansion of the gases of the swimming bladder that often led to the animal's demise. Studying other laboratory animals housed in pressure-controlled glass bells, Bert eventually described the quintessential physiologic changes that occur during rapid decompression. It was Paul Bert who thus finally identified the causative event for decompression sickness: nitrogen bubble formation.

After the acceptance of the theory of cell-based life proposed by Schleiden and Schwann (1839) and the publication of Darwin's *Origin of Species* (1859), scientists' view of life and its mysteries as the compilation of mathematical and microscopic events became all-pervasive. Michael

Figure 29  Paul Bert (1833–1886), the premier French physiologist of the nineteenth century, correctly identified nitrogen as the causative gas in the bends. An amateur politician as well, he served in the 3rd republic of France as minister of education and founded the Universities of Lyon and Lille. He died from dysentery during a diplomatic visit to Vietnam. (From Paul Bert, *Barometric Pressure*, Columbus: College Book Company, 1942)

Faraday's work with electromagnetism in the 1830s illustrated the physical laws of electricity and opened the doors to the submicroscopic world of James Clerk Maxwell (1831–1879). Maxwell's mathematical explanations for the behavior of light itself established quantum mechanics. Mendeleev's periodic table of the elements was designed by 1869 and showed the chemical relationships of all of earth's known atoms. The work of J. W. Gibbs (1839–1903) and James Joule (1819–1889) on the conservation of energy and thermodynamics in the 1870s tied together the behavior of atoms with heat, work, and machines. "It seemed for a while," writes J. D. Bernal, "as if the whole of natural phenomena could be explained in terms of simple observables of mechanical energy and heat."[3] With the understanding of ancient practices like fermentation, the field of microbiology was born. Louis Pasteur (1822–1895), a chemist and microbacteriologist, did more to establish medicine as a science in the fight against infection in the 1860s than any single figure up to that time. Although this was just before the advent of electricity, Mendelian genetics, quantum theory, antibiotics, and other advances of late nineteenth-century science, it was also a time of the first university-based research laboratories. Despite their academic associations they were partly, if not wholly directed toward some industrial, agrarian, or military application.[4] Paul Bert matured into this world, and his scientific work reflected the ever-increasing sophistication of the research climate around him.

In 1865 at the age of only thirty-two Bert was awarded a prestigious prize by the Académie des Sciences for his work on skin grafting.[5] He became interested in altitude physiology at a time when human balloon ascents were popular and when balloonists rose to heights above four miles without oxygen, armed only with courage, a pen, and a log book. In 1874, Bert sponsored the flight of three balloonists aboard the balloon *Zenith*, largely to outdo an earlier English ascent to twenty-four thousand feet in 1862 (fig. 30). At twenty-six thousand feet, all three men had become unconscious (two did not recover and died). Bert intended the balloonists to obtain objective, quantitative measurements of the upper atmosphere, but because of the cold and the rarefied air, no reliable data could be accumulated by either man or machine. Recovering from the shock of the accident and learning to pursue his scientific pursuits more assiduously, Bert moved his experiments indoors, relying on his strong background in the laboratory method.

In contrast to the gentleman scientists of Robert Boyle's day, with

Figure 30. Paul Bert sponsored this infamous voyage of the balloon *Zenith*. Reaching an altitude of twenty-six thousand feet, all three passengers became ill in the rarified air they had hoped to conquer; two died. This tragedy sent Bert permanently into the laboratory for the careful, quantitative animal experiments which would revolutionize baromedicine. (From Bert, *Barometric Pressure*)

the arrival of modern chemistry and then bacteriology and the natural sciences, experimental inquiries were largely performed by those in academia after long years spent in basic science studies and observation. By the mid-1800s, the laboratory method had become well-established. Older scientists tended to perform an experiment and then come up with a hypothesis to explain the findings (a posteriori reasoning). The "modern" scientists learned to first create a hypothesis, which is then tested by a carefully designed set of experiments (a priori reasoning). Data would be ana-

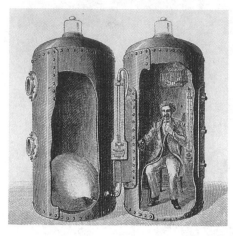

Figure 31. The compression chambers of Paul Bert. (From Bert, *Barometric Pressure*)

lyzed and the hypothesis confirmed into a working theory or discarded and the process started over.

Bert approached the problem of high altitude sickness the same way. His initial hypothesis was that low atmospheric pressure was certainly to blame, but he also wished to test the effect of the rate of *transition* to low atmospheric pressure. It had long been known, of course, that people at high elevations can easily tolerate life at fifteen thousand feet, whereas people accustomed to living at sea-level require several days to weeks to become acclimated to such heights. To test what occurs during this transition and how animals develop this adaptability, Bert constructed pressure chambers fashioned from bell jars (fig. 31). Access ports allowed him to obtain small quantities of venous or arterial blood at different times and at different pressures from as high as 10 atmospheres to as low as 0.1 atmosphere. His landmark studies on sudden decompression grew as an offshoot of this original project only by accident.

True genius, it is thought, lies not in succeeding but in seeing success in failure. Paul Bert was no different. Like Boyle two hundred years earlier, Bert first tested low pressure on various experimental animals, including sparrows, rabbits, dogs, cats, frogs, and carp. On one occasion, however, an otherwise healthy sparrow died soon after decompressing to low pressure too quickly, or so Bert surmised. An autopsy did not reveal much in the way of gross findings, so Bert tried increasing the pressure from which quick decompression was performed. Test animals for the experiments were usually housed in a chamber in which the air pressure was elevated 6–10 atmospheres. After about two hours of study, a stopcock was turned allowing the instant decompression of the chamber to 1 atmosphere. On October 27, 1870, Bert observed a sparrow thus decompressed from 12 atmospheres to 1, in seconds. As his notes described,

> Experiment #511 . . . When sudden decompression was made, darted to the top of the cylinder, then fell back. Was dead before being taken from the apparatus. Air in quantity in the jugulars and the right heart. . . .
>
> Experiment #519 May 24. Two rats . . . to 8 1/2 atmospheres; decompression in 2 minutes. The rats run about when taken from the apparatus; a few minutes after they become paralyzed and die. Gas is found in the whole venous system. . . .
>
> Experiment #528 [June 18, 1871]. Small dog . . . to 10 atmospheres; stays there about one hour; decompressed in 3 minutes. The animal cannot get out of the apparatus; there are no other movements than those of respiration; constant cries of pain. . . . At autopsy, . . . a cannula is passed into the right heart from which is extracted 33.9 cc of gas containing 20.8% of carbon dioxide and 79.2% of nitrogen, with some traces of oxygen. The right heart and the veins are full of gas and frothy blood; the same thing is true of the veins of the pia mater and the choroid plexuses.[6]

Ever the scientist, Bert took advantage of even the most horrific setbacks including the following accident in early 1872:

Experiment #608. Poodle placed in the apparatus in the morning. At 10:30, . . . it is well; the pressure is 9 1/2 atmospheres. . . . Immediately a violent explosion is heard. The porthole glass is burst and its fragments cut a lead water pipe a meter away. The apparatus was lifted, off from its supports and overthrown. I take out the animal with great difficulty for it has become cylindrical and is hard to pull through the door. Subcutaneous intra- and submuscular emphysema. Gas escapes whistling when the belly is opened. The right heart, as all the veins, is full of gas *but none in the left auricle or aorta*. The nerve fibres of the spinal cord are dissociated by bubbles of gas. . . . I extract 50 cc of gas from the right heart [which] contains: 1.9% oxygen; 15.1% carbon dioxide; 83.0% nitrogen."[7]

Almost two hundred years earlier, Robert Boyle had observed bubbles within the eye of a snake he had decompressed in his air pump. Inquiries as to the nature of the bubbles stopped there, as his work preceded the chemical revolution of the late eighteenth century. Paul Bert's laboratory, in contrast, benefited from the enormous outpouring of chemical and laboratory work in early and mid-nineteenth century, both in terms of knowledge as well as laboratory equipment and technical advances. Lavoisier and Cavendish had first determined that air was about 80 percent nitrogen and 20 percent oxygen as early as the 1780s, a fractional relationship that was constant anywhere scientists on the globe could obtain samples. Gas analysis was difficult, however, as no quantitative methods existed. Nitrogen, in fact, was particularly difficult to study as it did not react readily with other compounds with which to test its properties. This "chemical lifelessness" moved Lavoisier to originally refer to the isolated gas as "azote" or "without life" in the 1770s. However, by the 1850s, the German chemist Robert W. E. van Bunsen had developed methods of flame analysis with the burner that carries his name. He devised a test for individual gases by adding specific absorbents to a mixture of gases and then measuring the change in volume after the reaction. This process, called absorbent volumetry, was also joined by sophisticated glassware, valves, pipettes, pumps, barometers, audiometers, and flasks. By the 1860s, the gases in the blood could be routinely measured with absorbents, and it was soon recognized that blood served as the tissue in which oxygen is absorbed and carbon dioxide given off in the lungs.[8] Paul Bert was poised, therefore, to use seminal models of normal physiology to explain his observations of what was clearly abnormal.

He modified Bunsen's gas analysis method by constructing a mercury-filled pump that allowed blood extraction from a pressurized animal. The blood samples would be pumped and the evolved gases analyzed with the absorbent method. After the addition of potash, carbonic acid (carbon dioxide) was absorbed and then pyrogallic acid, which absorbed oxygen. The remaining gases, by inference, would be nitrogen. It is interesting, therefore, that Paul Bert, the father of compressed-air physiology, never actually measured nitrogen directly; this could not be accurately done until the advent of the spectrophotometer in the 1890s.[9]

Bert recalled the writings of Bucqyou and Rameoux of Strasbourg, Alsace-Lorraine, who in the early 1860s implicated the egress of dissolved blood gases after decompression in causing the emphysemas observed at Triger's mine at Douchy in 1845.[10] Bert's experiments showed, however, that "the gas which would threaten life on being liberated would be exclusively the one the proportion of which was considerably increased in the blood, that is, *nitrogen*."[11]

Bert believed that the small amounts of oxygen and carbonic acid relative to nitrogen obtained from the right hearts of decompressed animals were due to the free escape of these gases through the lungs. Nitrogen, however, does "not escape through the lungs because it is in an atmosphere which is four-fifths nitrogen and nothing urges it out."[12] Bert formulated his observations into theory:

The tissues of the organism . . . are laden with a growing proportion of nitrogen . . . and when decompression occurs, the gases must necessarily return to a free state, distending and even lacerating the tissues from which they escape. . . . The presence of such bubbles would be enough, I think, even if there were no stoppage of the circulation, to explain on the basis of irritation of the tissues, the slight symptoms of workmen in caissons . . . and be the cause of pains and local swellings when they do not cause death. We therefore understand the risks run by these workmen whose paralysis or death at these limits depends upon the size of a bubble of gas.[13]

Bert's identification of nitrogen as the causative gas in decompression sickness was a landmark event in bends research as well as general physiology. He could not, however, explain all aspects of the disease. He himself noted "another surprising thing," that of the "interval of 5 to . . . even 15 minutes which always elapses between the moment of decompression and that of paralysis."[14] He could not explain why one animal decompressed from a certain pressure would die, whereas a similar animal lived but with other symptoms.

It was the heterogeneity of the symptoms of decompression illness that had some physicians unconvinced of the nitrogen bubble hypothesis. Andrew Smith had at least heard of Bert's preliminary work during the completion of the Brooklyn tower of the East River Bridge. Bert's studies, as Smith read, involved animals decompressed from 10–19 atmospheres of pressure. How, Smith asked, could men suffer from nitrogen bubbles in the blood, which should be freely expirable through the lungs, when decompressing from caisson pressures as little as 2 atmospheres?[15] Smith brought a dog down into the Brooklyn caisson in the spring of 1871. After seven hours at 35 psi, the dog was killed and a venous sample of blood was placed in a flask capped by an oil-based air transducer. Finding only a small, benign fraction of air escaping upon decompression at the surface from the blood in the flask, Smith felt sure that release of gas in the blood was not the causative event.[16] The essential tissue, as researchers would show thirty years later, was not the blood but fat tissue. Had Smith analyzed the fatty tissue of the dog after decompression, he might have discovered the bubble-fat connection, now known to be crucial for the bends to occur, back in 1871.

Implicated in almost all theories of decompression sickness was Dalton's Law, which stated that the pressure exerted by a mixture of gases, in this case compressed air, is equal to the sum of the partial pressures of each gas.[17] There was little reason, therefore, to believe that nitrogen had any more proclivity toward bubble formation in the blood upon decompression than carbon dioxide or oxygen. However, James Henry, the nineteenth-century English chemist, had first postulated that the volume of a gas that could be dissolved in a liquid solution or tissue is proportional to the partial pressure of the gas. Given three gases at equal pressure—carbon dioxide, oxygen, and nitrogen—the amount dissolvable into a tissue should be proportional to each gas's partial pressure and its solubility in the tissue, or volume (V) = partial pressure (P) × solubility coefficient ($\alpha$).[18] Nitrogen, however, is five times more soluble in fat than any other atmospheric gas and, under compression, total body fat can absorb large amounts of the gas.[19] If one liter of oxygen is taken up by fat at a partial pressure of P, five liters of nitrogen at the same pressure P will be absorbed. According to Henry's Law, if the partial pressures of both gases are doubled, then ten liters of nitrogen will be absorbed and only two each of oxygen and carbon dioxide. Nitrogen is continuously absorbed into fatty tissue until the partial pressure of nitrogen in the blood equals the partial pressure in the body. When this point is reached, a body is saturated. Upon decompression, the blood partial pressure of nitrogen goes down very quickly and allows the body stores of nitrogen to be released. The greater the body store, the more nitrogen will need to be released to achieve equilibration. If the release is rapid enough, nitrogen bubbles can be formed. These then coalesce and arrive in the venous blood stream. At any point, from tissue to capillary to blood stream to the heart or to lungs, the true havoc of the bends can be wreaked.

Nitrogen absorption in tissue fat, therefore, is what predisposes an individual to decompression sickness. Smith himself noted that a high proportion of the 102 workers who suffered from the bends were "corpulent."[20] Eight of thirteen cases of paralysis and all deaths at the Brooklyn Bridge site occurred in what he termed "obese men." Men with more tissue fat, today easily measured by skin fold thickness, can accumulate a higher tissue load of dissolved nitrogen and thus a greater risk of developing decompression sickness.[21] It would be left to the even more careful scientific inquiry of the twentieth century to determine how to overcome the condition.

# 10

## John Scott Haldane and Staged Decompression

Throughout the nineteenth century, clues existed as to why the nitrogen-fat connection was the causative relationship in decompression sickness, perhaps first noted by Andrew Smith during the Brooklyn Bridge project in the 1870s. John Scott Haldane (1860–1936), scion of a noted Scottish family of soldiers, explorers, thinkers, and scientists, took the analysis to the in vivo level, which led to the development of accurate decompression tables (fig. 32).

Haldane became interested in gas physiology as a schoolboy, and first reported the effects of non-circulated air in the men's rooms of his first boarding school at Sussex in 1879. Haldane's interest in air and environmental disease led him in adulthood to investigate mining accidents involving carbon monoxide in coal miners.[1] Haldane developed sophisticated and sensitive gas detectors that allowed him to first demonstrate how carbon monoxide and oxygen bind to hemoglobin. He took advantage of the reversible equation describing the displacement of covalently bound carbon monoxide from hemoglobin by high concentrations of oxygen to develop a treatment for patients suffering from carbon monoxide poisoning.[2] Haldane first administered pure oxygen for victims of carbon monoxide inhalation in 1919, and oxygen is still a mainstay of emergency room therapy for poi-

Figure 32. John Scott Haldane developed the concept of "stage decompression" and wrote the first truly modern decompression tables. Versions of his schedules are still used by navies today. (Lafayette Ltd., London)

soned miners, firefighters, and people who have been exposed to car exhaust.[3] Haldane's work established safety codes for mine air composition and routine carbon monoxide testing, thus solidifying his reputation as an advocate for occupational reform.

Lavoisier and later Spallanzani had furthered Joseph Priestley's work and identified oxygen and carbon dioxide by the end of eighteenth century as the "vital" gases necessary to sustain life.[4] Haldane and his associate J. G. Priestley in 1905 finally demonstrated the importance of blood carbon dioxide (and therefore blood pH) in determining the medullary control of respiration in the brain stem. The higher the acid content (the lower the pH), the more active is the breathing center of the brainstem. As

Haldane showed, it is the blood pH and not carbon dioxide directly, therefore, which is responsible for our subconscious impulses to respire. His work was also seminal in understanding tissue gas physiology. Physiologists throughout the world still refer to the Haldane effect: the ability of red blood cells to absorb waste carbon dioxide in times of decreased oxygenation or increased acidity in the bloodstream. Such achievements remain a testimony to Haldane's lasting impression in the world of biology and our understanding of how molecules and atoms interact within the body.

As with many advances in medicine and science, military goals fueled research on the effects of decompression. The Royal Navy in particular had a great need to advance its use of compressed air for diving, salvaging, and repairing at sea but had no details about safe decompression rules. Haldane's colleague Lieutenant Guybon Damant was not only on staff at Guy's Hospital in London but was also a naval officer and sitting member of the Admiralty Committee on Deep Diving and had worked on earlier diving tests off the coast of Scotland. Haldane's repute in gas physiology, even by the early 1900s, placed him at the forefront for any advanced laboratory experimentation with compressed air. His colleagues Dr. A. E. Boycott and Damant would be vital in these projects.

Haldane was long an enthusiastic follower of the work of Paul Bert. Haldane gave much credit to Bert for earlier recognizing the qualitative effects of oxygen depletion and low atmospheric pressure on the body in vivo and for recognizing nitrogen as the cause of decompression sickness. Haldane and his colleagues knew that Bert's work on decompression sickness contained the greatest source of initial data from which to plan further experiments. It was now up to Haldane and colleagues to determine how to control the nitrogen gas and establish rules to make breathing compressed air safe at any depth for any given length of time.

However, a more controlled compression chamber would have to be built. With the help of the Navy and the Lister Institute of

Preventive Medicine a large steel chamber was provided at the Institute's compound in London. Definitely not recommended for claustrophobics, the chamber was essentially a boiler seven and a half feet long, resting on its side with two six-inch windows on each end, a single two-foot manhole entrance, and a six-inch lock for passage of food. In it, goats, sheep, or human beings could comfortably habitate for any given amount of time with pressure production and release governed by a complex series of valves and controls operated by a round-the-clock research crew.

Haldane had read H. M. Vernon's work in 1907 on the high solubility of gases in oils and recalled that, at body temperature, nitrogen solubility was five times greater in fat than in water. "The proportion of fat and fatty material" he wrote with Boycott, "is very different in different parts of the body, so that the capacities of different tissues for taking up nitrogen must vary accordingly."[5] The workers focused their efforts on all the different tissues of the body, their rates of gaseous uptake, and the rates at which bubbles formed under different pressures and decompression rates.

What emerged in victim reports and observed animals was that the nervous system was most severely affected during bends attacks. The excruciating paralysis, limb numbness, and nerve damage were hallmarks of the decompression disease. Spinal cord sections from goats acutely decompressed in the Lister Institute chamber were painstakingly analyzed. Robert Boyle had been the first to see air bubbles in the eye of a snake he had acutely decompressed in the 1680s, and Bert in the 1880s identified bubbles in the right side of the hearts of animals killed by acute or sudden decompression. Haldane examined nervous tissues under the microscope with then-current histologic and staining techniques. Extravascular bubbles were confined to the white matter (fatty nerve tissue) and the zones where circulation was relatively poor. The myelin sheaths of nerve fibers, because of a relatively high lipid (fat) content, formed "reservoirs of dissolved nitrogen."[6]

Haldane wrote, "For this reason, nitrogen will tend to be liberated in the white matter of the brain and spinal cord . . . and large nerves. . . . The bends and other symptoms from which workers in compressed air so frequently suffer are probably due to liberation of bubbles from the gas dissolved in myelin sheaths."[7]

Dissolving whole rats in strong potash (potassium hydroxide), Haldane analyzed total animal fatty acids and found that animals which died after decompression had 25 percent more fat than survivors. The results indicated that "mortality runs parallel with fatness rather than with size."[8] "The practical conclusions are clear," Haldane wrote. "Really fat men should never be allowed to work in compressed air and plump men should be excluded from high pressure caissons or in diving to more than 10 fathoms. . . . It is unfortunate that an increase of experience and skill in technical operations should so often be associated with the increase in waist measurement which accompanies the onset of middle life."[9] Such observations had been anecdotally and informally noted by almost all physicians assigned to compressed-air work projects—Pol and Watelle, François, Jaminet, and Smith all had noted most deaths occurring in those unhealthy men who were fat—and form part of the criterion for U.S. Navy divers today. It was first stated in the manual of the Navy's medical department of 1958 that diving "candidates should present no greater than 10 percent variation from standard age-height-weight tables . . . [unless] due to heavy bone and muscular structures."[10]

As the *cause* of decompression sickness became better understood, the treatment and *prevention* of it remained largely unstudied. Once a worker became afflicted with the bends, there was still little that a physician at the turn of the century could provide. Some workers noticed an immediate cessation of arm pain if they plunged the afflicted limb into a pail of liquid mercury. This was later thought to increase the hydrostatic pressure on the affected limb and cause bubble formation to cease or decrease. Pol and Watelle had recommended total body recompression as early as the 1840s for bends sufferers but never actually went through with the idea. Paul Bert himself had noted the curative effect of recompression on his experimental animals. Medical recompression chambers, suggested by Pol and Watelle in the 1850s, New York's Andrew Smith in 1881, and actually built by Ernest Moir for the Hudson River tunnel in New York in 1889, could ameliorate bends symptoms but could not prevent them. Smith's and Jaminet's decompression tables in the American bridge projects were the first for humans, but they were purely empiric and entirely unreliable in preventing the bends for varying pressures. Up to 1908, there was no way to accurately predict the human response to decompression from a given pressure experienced for any given time. Haldane's work was the basis from which such a modern system would be built.

Haldane and his colleagues continued hundreds of trials of tests on animals and humans from 1903 to 1904 (including Haldane's twelve-year-old son, the future anthropologist and geneticist J. B. S. Haldane), taking careful measurements of blood gas concentrations and symptoms. Haldane noted that at 1 atmosphere of pressure the air has a total pressure of 713 mm Hg, excluding water vapor pressure, made up of the partial pressures of nitrogen (586.45 mm Hg), oxygen (158.25 mm Hg), and carbon dioxide (0.30 mm Hg). With increased pressure, the partial pressure contributions of each gas remained the same and Dalton's Law was preserved (at 5 atmospheres, that is, the pressure of nitrogen was still about 80 percent of the total). At increased pressures, the gradient for nitrogen and oxygen absorption were proportionally increased. Carbon dioxide, which has a very small contribution to the total pressure, also increased proportionally but was still much lower than venous carbon dioxide partial pressures (46 mm Hg), allowing the unimpeded delivery of the gas to the alveoli.

With an increased gradient, each time blood passes through the lungs it picks up more and more nitrogen and delivers it to the tissues of the body. With each pass, nitrogen is absorbed by all tissues, and according to Henry's Law, the amount absorbed is proportional to its partial pressure. Some tissues have better blood flow and they saturate faster than tissues with poorer blood flow. Watery tissues (like muscles) saturate faster than fatty tissues (such as nerves). However, Haldane recalled that fatty tissues can absorb five times more nitrogen than other tissues, so fatty tissues with poor blood flow took the longest to saturate.

Released from high pressure, the reverse situation occurred and nitrogen "came out" of its absorbed state at rates proportional to the blood flow of the tissue as well as the tissue's fat content. The fattier tissues with poorer blood flow were thus the last to desaturate. Haldane also believed that tissues could become "supersaturated" with nitrogen after decompression. This implied that after the release of pressure, bubbles were not formed right away and that some tissues could tolerate this supersaturated state. (We now know that even after minor decompression, from three to six *feet*, very small bubbles can be formed, but they do not cause symptoms.) Determining the pressures at which Haldane's supersaturated states occurred led to his modern decompression theories.

Haldane's contributions to hyperbaric medicine and diving physiology thus stemmed from two essential observations. First, he found that the severity of decompression symptoms increased proportionally in subjects with time until a plateau was reached at about three hours of exposure to compressed air. This was evidence for a point at which the body's reservoirs became completely saturated with nitrogen (about twelve hours in humans) after which further uptake cannot occur: the state of nitrogen saturation. (Oxygen absorption is limited by the amount of oxygen that can be covalently bound to hemoglobin and other proteins so the point of total oxygen saturation occurs much earlier and at a lower oxygen partial pressure.) Second, the longer a tissue was allowed to saturate and the more nitrogen it absorbed, the longer it took to desaturate the tissue before a critical level at which bubbles formed. The amount of nitrogen absorbed by a given tissue was dependent on the fat content and blood volume circulating through it. This led to Haldane's concept of a range of tissue deposition of nitrogen in the body, the release of which would be heterogeneous over time. Uniform decompression, as had been the rule, was therefore shown by Haldane to be particularly dangerous if it happened too fast because it did not account for tissues like fatty nerve tissues of the brain or spinal cord which released nitrogen more slowly: "This slowness has never hitherto been recognized but must evidently be reckoned with in devising measures for the prevention of caisson disease. . . . It is needlessly slow at the beginning and usually dangerously quick at the end. . . . On the other hand, decompression should . . . be as rapid as possible, consistently with safety." [11]

Haldane observed that no case of the bends had occurred following a decompression from less than 1.25 atmospheres. Indeed, 1.6 atmospheres was the lowest pressure recorded for a decompression fatality.[12] Haldane believed, as was shown in the Lister tank, that if the levels of supersaturation could be calculated, it would be safe to go from a compressed state to a decompressed state *up to the point where supersaturation was no longer tolerated* and bubbles occurred.[13] When decompressing from a known pressure of P to, for example, one-half P, the volume of gas liberated is the same regardless of the magnitude of P. It now appears that very tiny bubbles form at even minor drops in pressure but never cause a problem. Haldane's theory, which was based on experimental data and derivations therefrom, may have been incorrect as regarding supersaturation, but this did not change his numbers and therefore his mathematical approach to decompression. He concluded, "Hence, if it is safe to decompress suddenly from two atmospheres to one, it would be equally safe to decompress from six to three, etc. Our experiments have shown that this is the case." [14]

This gave rise to Haldane's theory of tissue *half-times*—that the likelihood of developing bubbles to cause decompression illness was proportional to the amount of nitrogen dissolved in the tissue with the slowest ability to desaturate to a nitrogen tension that was *one-half* of its supersaturated state. Fatty tissues have, therefore, a longer half-time than more watery tissues like muscle.

Using this line of inquiry, Haldane and coworkers devised a rapid method of decompression which later became known as "stage decompression" of half-time tissues. A diver or worker is decompressed rapidly to one half the pressure at which he worked, and *slowly* decompressed thereafter. Decompressive experiments on Haldane's colleagues, and his own son, from pressures as high as 6.4 atmospheres or almost 80 psig, showed the hypothesis of stage decompression to hold true.

After a dive to 213 feet (7.5 atmospheres) for a long enough time to achieve total body saturation (several hours), the diver could rapidly ascent to 90 feet (3.75 atmospheres) for the first stage. The next stages, calculated by Haldane, were slow, allowing for equilibration of nitrogen in the blood with inhaled compressed air at each depth so that "the nitrogen pressure in no part of the body ever becomes more than twice that of air." [15] Eventually, after decompression stages at ten-foot intervals and five hours later, the diver could surface. Haldane did not calculate the exact total supersaturation levels for a particular subject's dive depth or time, but overestimated the time it took to decompress from it. This resulted in safer decompression overall but sometimes faster as well, which Haldane readily pointed out. Today's diving tables have had many empiric adjustments through the years, but all incorporate the important first stage ascent on the way to the surface. Divers have learned where precious minutes can be shaved and ascents quickened. Correspondingly, for longer, deeper dives, for which Haldane may have undercalculated, extra times and stops have been added by Navy divers. In the U.S. Navy tables from the early 1950s, for example, seven or eight extra stops were added to the 213-foot dive described above, for a total ascent time of fifteen and a half hours.[16]

Because such a long time in water can quickly lead to fatal hypothermia, decompression in the water may be impractical. Haldane's calculations, in fact, showed that depending on a diver's depth, he or she could go all the way to the surface in one long ascent if they could get into a recompression chamber quickly and "dive" back to a "safe" depth for a slower, staged decompression. This relies on Haldane's previously described theory on the body's ability to tolerate a state of "supersaturation" before bubbles are formed (or before bubbles of troublesome size are formed) after acute drops in ambient pressure to 1 atmosphere. Such decompressions are today completed on the surface in specialized recompression chambers first developed before World War I by Sir Robert Davis.

It is now accepted that decompressions from even one *foot* cause "microbubbles," but they are harmless and seem to be tolerated well by the body.[17] At a certain point, however, bubble size does become a problem. Haldane correctly observed that *most* divers can tolerate a decompression from 2X atmospheres to 1X atmospheres without symptoms. This, it turns out, was due not to Haldane's elegant supersaturated tissue hypothesis but simply to the size of bubbles generated. Still, Haldane's invention of staged decompression based on these observations and his designation of tissue half-times were vital to the development of modern decompression theories and diving tables.

After Haldane and his laboratory colleagues had presented their work to the British Admiralty Deep Diving Committee, the scientists, confident of their own mathematics and presence of mind, tried out Haldane's staged decompressions in the open water on themselves. Lieutenant Guybon Damant found "going underwater a delightful experience and infinitely [preferable] to the study of ballistics and gun drills."[18] Damant and gunner Andrew Catto dived from the H.M.S. *Spanker* in 1906 first to 138 feet. Following Haldane's staged method, they surfaced intact. Then, they dived to 180 feet and, finally, 200 feet. Each time, the two divers surfaced without any complaints. For still deeper, and more dangerously frigid waters, the *Spanker* was sailed into Scotland's Loch Striven where the two reached the bottom, 210 feet down, where they tolerated pressures of 7 ATA (85 psig). Ascending first to about 100 feet, the two slowly went by stages to the surface, where they greeted their cheering colleagues after three hours, only a few minutes of which was spent at depth.

Haldane's staged ascents were usually shorter than those with uniform decompression, but the stays underwater were still quite long, limiting active working time, and were completely dependent on good weather and calm seas. As mercantile traffic and shipwreck salvage projects increased in the early twentieth century, deeper dives required decompression times that became impractically long. To aid in this regard, a decompression bell was designed and built by Sir Robert Davis of Siebe, Gorman & Company in 1910 and was submerged to the level of the first Haldanian ascent stage for a given dive depth (fig. 33). An attendant inside the bell assisted the diver in entering, and the bell was then brought to the surface for the remaining stages of decompression, with the diver comfortable and warm. The bell was routinely used through World War I and was used in the salvage of the submarine H.M.S. *Poseidon* in 1931 in the South China Sea. These decompression bells evolved into modern-day decompression "quarters" or deck decompression chambers in which divers can comfortably relax, eat, obtain medical attention, and sleep as they slowly "surface." [19] For long dives of hours at depths of hundreds of feet, decompression at the surface can take days, and divers can while away the hours in comfort as they "ascend" in the chamber to 1 atmosphere.

Few divers could use Haldane's schedules in 1910. Even by 1925, only twenty U.S. Navy divers were qualified to dive to

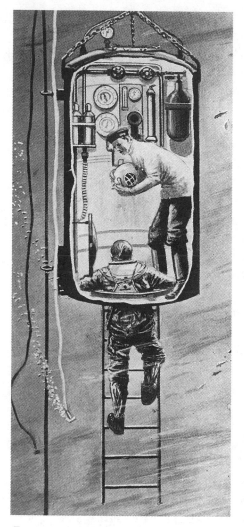

Figure 33. The decompression bell by Sir Robert Davis developed in the early 1900s. Submerged at the first Haldanian stage, an attendant awaits the ascending diver. The bell allowed a warm environ to decompress safely and could be brought to the surface whenever the sea proved too rough. (Reproduced from Haldane and Priestley, *Respiration*, New Haven: Yale University Press, 1922)

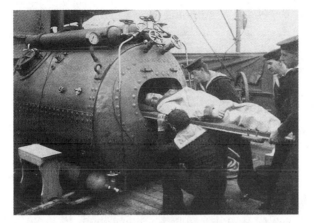

Figure 34. Decompression chamber, World War I era. The chamber was every bit as claustrophobic as it appears: divers would be locked tightly into the chamber as they were quickly recompressed and then decompressed over several hours. (Reproduced from Haldane and Priestley, *Respiration*)

90 feet, and only six civilian divers claimed to be able to reach much deeper than 130 feet.[20] The most immediate application of such sound laboratory based knowledge *should* have been in that other strange world of human accomplishment in which compressed air was the secret to success: tunneling. For reasons that remain unclear, American tunneling firms did not follow Haldanian principles for decompression—with terrifying consequences—whereas British tunneling firms gradually enacted staged decompressions into their manuals. During the next twenty years there was not a unification of Haldane's diving theories with the practical experience of American engineers who use compressed air but a divergence; instead of a standardization of decompression tables and recompression medical care, there was a promulgation of substandard rules; instead of a peer-reviewed process of scientific discourse and engineering debate, there was an untested system of care that still requires major modifications.

Diving and engineering became functionally separate worlds of compressed air use. Haldane, although first interested in the atmospheric problems in mine work, was paid by the Royal Navy to do the research that eventually lead him to discover staged decompression and Haldanian decompression tables. These tables were immediately employed, therefore, by naval physicians and divers by 1910 and used by the maritime community, thus establishing a generation of naval and commercial compressed-air research based on some aspect of staged decompression (fig. 34).

The engineering world, however, had no Haldane. Although engineers and caisson designers had easy access to the diving literature, Haldanian tables, and decompression technology, the tunneling field seemed to differ from the maritime field in how decompression sickness was handled. Original compressed-air work decompression tables for engineers did not adhere to the Haldanian strategy of staged decompression. By the 1920s, caisson workers and naval divers adhered to completely different schedules for decompression. Compressed-air work in a tunnel is often much longer (causing more nitrogen saturation) and more strenuous and decompressions are more frequent than those to which naval divers are routinely subjected. Also, no professional diver would voluntarily inhale compressed air containing the gas pollutants and exhaust prevalent in the average tunnel shaft or caisson. It is somewhat amazing, therefore, that there was little if any collaborative effort between writers of decompression tables for tunneling workers and those for navies. This divergence has continued to this day as the naval and private commercial diving firms having continued to develop ultramodern decompression techniques while tunneling and caisson workers still follow outmoded timetables. Understanding how this came about illustrates the chasms that can easily develop in our technological world and may help guide us back to a more unified community of engineers, scientists, and the people who benefit from their work.

# 11

# Decompression Sickness

# and the Government

Government interest in workers' health is fairly recent,
stemming partly from the laissez-faire mentality that still
pervades modern capitalism, as well as the government's
inability to enforce rules or regulations effectively. The
evolving struggle to preserve workers' health and yet
maintain employers' economic success has become so en-
trenched in today's socioeconomic thinking that it seems
markedly odd that employee health-related laws were
not routinely advocated until the end of the nineteenth
century. Due in part to the inherent disorganization and
voicelessness of the poor during most of the Industrial
Revolution as well as a profound insensitivity of the era's
employers, there was only sporadic, albeit violent, reaction by the populace to the status quo. The Parisian up-
rising of the Commune in the 1870s was a harbinger of
the growing hurricane of economic unrest.[1] Soon, the
entire Euro-Asian economic axis of power would be un-
seated, first with the Bolshevik revolution in the 1900s.
Political efforts to improve worker health by the 1900s
may have been due less to a legitimate desire to help the
daily lot of a country's laborers than a desire to stem the
tide of worker revolt and nationwide economic instability.

Compressed air may have been something new by the

1840s and 1850s, but the pervasive industrial disregard for other substances threatened to make its use no less hazardous to laborers than factory smoke, arsenic, mercury, coke, and coal over the previous fifty years.[2] Bridge builders James Eads and the Roeblings were remarkable not only for their pioneering engineering efforts but for relatively progressive views of their employees' health. They appreciated that their use of compressed air threatened disaster even as it promised success. Caisson work was unique, for despite the fact that compressed air was invisible, the hazards associated with the work were strikingly apparent to both laborer and employer. The caissons were cramped, chokingly humid, and poorly illuminated. Unrest was easily fomented. The difficult conditions, the constant threat of a watery death, and the image of thousands of tons of rock bearing down ten feet above one's head sustained a strong esprit de corps as well as a self-checked suspicion of those on the surface.

Eads and Washington Roebling knew well the crippling effects of strikes and worker dissatisfaction. With the sheer size of their projects, however, they were under a constant public scrutiny that no industrialist, engineer, or employer had experienced. Daily to weekly reports were published in the St. Louis and New York papers. During the 1870s and 1880s the evolving beauty of the Brooklyn Bridge graced the covers of *Harper's Weekly*, the *New York Herald*, and *Leslie's* newsmagazines no less than eight times.[3] So much interest in the goings-on "down in the caisson" was voiced by the American amateur scientific community that editors for *Scientific American* sent down a staff artist and lithographer to record the now-famous images of the excavation in the Brooklyn caisson during the 1871–72 season.[4]

With any project of such scale, there would be injuries and even deaths. But with compressed air, no explanations could be given and no preventive measures provided to the public to allay fears that workers' safety could not be maintained or, worse, that with each foot of digging, death rates would increase prohibitively. Had enough workers become ill or died from caisson work or the bad press become overwhelming, the projects could be derailed or scrapped altogether. It is unfair to assume that Eads or Roebling hired their personal physicians to care for their men for reasons of engineering and economic survival. But regardless of their aims, the decision to hire bona fide medical care on-site for employee health was clearly ten if not fifty years ahead of its time. The factories in Lowell, Massachusetts, regarded as the first in the world to have their own medical facilities, were built twenty years *after* the last caisson worker "locked out" from Roebling's bridge. The physicians for Roebling and Eads, Andrew Smith and Alphonse Jaminet, respectively, began an attempt by the world's engineering and medical community to understand why compressed air is hazardous, how to decompress from it, and how to predict a safe decompression schedule for any individual from any pressure and work time length. A hundred years later, that undertaking continues.

Smith and Jaminet were the first American physicians who attempted to understand the maladies of industrial accidents in the past to prevent those of the future. Such efforts preceded by fifty years the creation of the American Industrial Hygiene Association in 1939, which formalized industrial medicine as a separate discipline. Lacking sophisticated scientific equipment or the training to derive objective data, nineteenth-century physicians are important not for their descriptions of the causes of decompression sickness but for their attempts to establish decompression tables. These tables, although profoundly inadequate, established an industrial minimum so that any future projects using compressed air must also follow standard guidelines of decompression as well. The air-lock tender would no longer be allowed to open or close gas valves according to his or his workers' schedules but rather the schedule as mandated by the engineers and physicians. These schedules evolved through a collective effort between engineers (who wanted to minimize the time required for decompression), physicians (who wanted to minimize the suffering from the bends), and the government (who wanted to maintain an ability to suspend work if "minimum standards" were not established). The scientific community provided the data upon which compressed air regulations could be legislated as early as 1905. It would be another decade before such laws would be enacted. As is often the case, landmark legislative decisions often follow public outcry after accidents or tragic events. The majority of advances in industrial regulation in the world largely took place in Great Britain with the Factory Acts of 1802, the 1830s, and a major renewal in 1867.[5] These dealt mostly with hours of work and child labor. None dealt with worker health until 1898, when Parliament assigned Sir Thomas Legge as the first government medical inspector in the world.[6] The delay in this seemingly important appointment was due to an almost complete ignorance of worker health in the medical community. Separate medical disciplines to care for sick or injured workers did not yet exist.

Bernardino Ramazzini's (1633–1714) classic *De Morbus Artificium Diatriba* (Diseases of workers) in 1700 was revolutionary in that the treatise was the first to demand that physicians ask their patients about occupational hazards. Ramazzini himself remarked that "we must admit that the workers in certain arts and crafts sometimes derive from them grave injuries, so that where they hoped for a subsistence that would prolong their lives and feed their families, they are too often repaid with the most dangerous diseases and finally, uttering curses on the profession to which they had devoted themselves, they desert their post among the living."[7] His aim was to "incite others to lend a helping hand until we obtain a really complete and thorough treatise worthy of a place in . . . medicine [and] owe this to the wretched condition of the workers from whose manual toil, so necessary though sometimes very mean and sordid, so many benefits accrue to the commonwealth of mankind; yes, this debt must be paid by the most glorious of all the arts, as Hippocrates calls it in his Precepts, that of Medicine, 'which cures without a fee and succors the poor.'"[8]

No single physician advanced the public policy for the care of the occupationally ill more than England's Charles Turner Thackrah, and his 1832 treatise on the subject created occupational medicine. Thackrah recognized early the brutality of the factory system and the disregard of the employer for his employees' health. "Acquainted far less with physiology than with political economy their better feelings will be overcome by profit and they will reason themselves into the belief that the employment is by no means so unhealthy as some persons pretend."[9] "Miners," Thackrah remarked, "take immense quantities of ardent spirits, not with the view of enabling them the better to sustain their unhealthy employment but confessedly to drown the ever-recurring idea, that they are, from their occupation, doomed to premature disease."[10]

Inspector Legge's efforts finally led to the Factory Act of 1901, when a physical exam of all labor applicants, like that required by Jaminet, Smith, and Moir in America forty years before, became English law.[11] The eight-hour work day was established by Parliament in 1909 with the great Miner Act and on-site rescue personnel provided the following year.[12] Four decades after Jaminet and Smith's employee clinics, the British Police Factories Act of 1916 compelled employers to provide workers with clothing, first aid, washing and dressing facilities, and the preparation of meals.[13] Although Great Britain made significant industrial health progress during this time, not until the infamous Triangle Shirt Factory fire of 1911 in New York did any significant legislative regulations arise in the United States. Just the previous year, the first Conference on Industrial Disease convened in 1910. The first United States agency committed to worker policy was the Bureau of Mines, established by Congress to maintain safety and mining standards in the coal and metal industry.

Even the Depression could not slow nationwide efforts by physicians to improve industrial standards. By 1938 the American Conference of Governmental Industrial Hygienists was convened, eighteen years after Dr. W. G. Thompson's clinic for occupational diseases opened at Cornell University. Representing a large and diverse group of public health advocates, physicians, labor activists, and health care providers, the American Industrial Hygiene Association was formed in 1939. With the labor movement of the 1940s and increased federal commitment to occupational health research and training, worker compensation laws were passed swiftly by all contiguous states by 1948. To administer the multitude of federal, state, and local regulations regarding occupational and environmental disease, Congress established the Occupational Safety and Health Administration (OSHA) within the Department of Labor to oversee all facets of industrial health and hygiene, prevention, and cure, and in 1971, President Richard Nixon signed into law the Occupational Safety and Health Act. In its ideal form, OSHA was intended to identify and improve work-related health hazards, educate employers and their employees to establish maximum safety protocols, and hire organizations to conduct research in controversial areas of industrial hygiene and science. The research branch, the National Institute for Occupational Safety and Health (NIOSH), was the source of technical advice and scientific findings for OSHA.

For compressed air safety and maritime standards, OSHA promised to provide what had been needed for sixty years: a universalized approach to decompression sickness and its treatment. Unfortunately, as an illustration to the inefficiencies and difficulties in any large federal bureaucracy, even one with the noblest of goals, such seemingly reachable aims remain unmet.

While many fields of industry, including mining, textiles, agriculture, and farming were closely studied and improved, the tunneling community seemed to be escaping public and bureaucratic scrutiny. One principle that did appear to pervade the thinking of almost all tunneling engineers was that the likelihood of a bends accident occurring was clearly related to the total exposure time in compressed air. This made sense even to Andrew Smith in 1880 because those who made quick trips into a caisson, like Roebling or other engineers, had few or no symptoms when decompressing, compared to workers who had spent eight to ten hours at 20 or 30 psig. Eventually, engineers for other tunneling projects of major urban centers had established a simplistic version of minimizing both exposure time and delays in the work shift. If an eight-hour shift was dangerously long, than perhaps breaking it up into *two* shifts might allow all of the gas to escape from a worker's tissues so he could begin the next shift "fresh." The split shift, as it came to be known, was first used in the great 2,300-foot long underwater section of St. Clair tunnel under the St. Lawrence River where, in addition, decompression times were lengthened.[14]

By the time the Pennsylvania Railroad Company began construction of its Hudson and East River tunnel systems in 1904, the dangers of compressed air and the importance of medical recompression were clearly established. The Haldanian approach to decompression in stages was established by 1907 just in time for a great increase in tunneling activity throughout the world, especially New York City, where two large tunnels for railway and one for gas lines were built in 1909–1913 alone. For reasons that are unclear, few engineers in the United States put much merit in Haldanian stages, probably because it required longer stays for longer work shifts. Instead, uniform decompression was carried out at both ends of the split shift, which most workers saw as an imposition rather than as an attempt to preserve their health.

By 1902, every tunneling project usually had a medical officer on site, but the method of decompression was by no means standard and such differences were prevalent throughout Europe. An Austrian tunneler in Vienna would, for example, work at 3 ATA for two hours and be decompressed for thirty minutes. In Paris, his decompression time would be forty-five minutes. A job in Holland at the same time and pressure would require him to sit in a decompression chamber for sixty minutes.[15]

As the tunnels of the Pennsylvania Railroad crept their way under the mud of the Hudson between 1904 and 1909, the distance to the decompression units increased. To remedy this, the engineers had installed three separate compartments of pressure so the entire tunnel did not have to be pressurized at all times. One segment was kept at 40 psig, a middle one at 29 psig, and a third at 12.5.[16] Therefore, a worker had to *walk* through all three pressure decrements without regard to the now-appreciated physiologic strain on bubble formation. For the workman after *each shift* who desired a lunch on the shore, five minutes were spent in the first then, after a thousand-foot walk, eight minutes in the middle, and then fifteen for the last. After twenty minutes of walking, the worker spent only forty minutes decompressing from as high as 40 psig—a practice recognized today as extremely dangerous.[17] By the day's standards, most thought they did quite well. In 3,692 cases of the bends, twenty men died.[18] No laboratory or out-of-workplace studies were done to determine the best way for workers to maximize work but minimize physical harm. These tables allowed far too short decompression times, as Haldane would demonstrate, at higher pressures and longer shift lengths, and were needlessly long for shorter submersions.

Across the ocean, almost all the European nations had advanced compressed-air projects underway, and each followed its own system for decompression. Germans produced three tables between 1878 and 1899; the most conservative ones by Heller, Mager, and von Schrotter were ultimately accepted for widespread use. These early decompression-sickness theorists formulated their uniform, non-Haldanian tables by field observations of workers at tunnel excavations under the Danube at Nussdorf in 1899.[19] The British used their own tables for the building of the Newcastle-on-Tyne bridge in 1905–06 by the Cleveland Bridge and Engineering Company.[20] Eventually the French Ministry of Labor finally enforced decompression schedules by the Commission of Industrial Hygiene by 1906, and these became the first laws in the world to require adherence to specific decompression rates. These tables not only stipulated how fast a uniform decompression for a specific pressure would be undertaken but limited the shift lengths as well. However, several critics, including J. P. Langlois, pointed out that French law prohibited the regulation of hours of labor.[21] It is unclear if this was the reason for the tremendous noncompliance with the law among French construction firms. Workers were still following the pervasive attitudes that their work risks were their own and a complaint of the bends was a threat to their bringing home the day's ten francs. Eventually, the Austrians followed a uniform decompression rate of one and one-half pounds every two minutes, the French required a pressure-based rate, and the Dutch adhered to even slower decompression rates.

With the building of the East River component of the Pennsylvania Railroad tunnel system around Manhattan in 1904–09, Henry Japp and his medical officer, F. L. Keays, faced the formidable challenge of deciding which decompression system to use: classic Haldanian staged decompression or uniform decompression and at which rates and times. Keays felt that one of the dangers of the split shift was that it exposed the worker to two decompressions and thus twice the inherent dangers. With constant pressure by the labor leaders of the Air Worker's Union to maintain the split shift to increase the payable working time, Keays and Japp used a modified Haldanian system (staged decompressions) only if medical recompression was required. Otherwise, all usual decompressions were uniform. A man with the bends would be recompressed by Dr. Keays until he felt no pain and then was rapidly decompressed to 10 or 15 psig. The second decompression would then be uniform, slow, and up to the judgment of the medical officer. The Keays-Japp tables, used between

1906 and 1909, decreased the shift lengths with increasing pressures but by using staged methods were able to decrease the decompression times as well. The final Keays-Japp recommendations were a compromise among the medical, engineering, and labor worlds and therefore satisfied nobody entirely. Still, they were reasonably successful in decreasing bends accidents during the East River Tunnel construction. In a half million decompressions between 1904 and 1909, twenty fatal cases occurred up to 42 psig. After the Keays-Japp tables were in use, no "severe or fatal cases occurred and such cases as were treated were slight."[22]

Perhaps the most unhappy group regarding decompression was the workers themselves. Either lacking the sophistication of their engineering or medical supervisors regarding the hazardous potential of compressed air or filled with the bravado and esprit de corps that is usually seen in those who routinely deal with dangers, workers or "sandhogs" saw decompression as a nuisance and a complete waste of time. Often, men would try to "escape" decompression altogether by using a supplies lock to reach the surface and decompress to sea level (1 ATA) from pressures as high as 4 ATA.[23] Such disregard for medical precautions and personal neglect in caisson work were noted sadly by Jaminet and Smith in the 1870s. Those that did follow orders did so without ever understanding why exactly they had to spend hours in a dark, iron air lock, sitting crammed on hard benches without a thing to do. Some would drink "tunnel coffee," a thick, strong brew. One sandhog brought in a bottle of carbonated ginger ale, but under pressure in the caisson the drink went completely flat. Thinking it truly without fizz, he discarded it, but another worker retrieved it and drank half of it. At the end of the shift, the group decompressed together and, with lower pressures, the inherent carbonation of the ginger ale made its presence known dramatically. The cork popped off the bottle with such a loudness that one sandhog thought he had been shot. The less fortunate victim, of course, was the chap who drank the half bottle of "flat" soda only to find his stomach and intestines now filled with gas.[24] Those in the air lock with him at the time apparently shared in his misery.

Facing pressure from New York Governor John Williams, the Industrial Commissioner helped pass the first state regulations of compressed air use on May 6, 1909, based largely on the experience of Japp and Keays. The law, the first in the United States regarding pressurized air, required a medical officer to attend the locks at all times and specified maximum shift lengths, physical exam standards, and penalties for violations. A ceiling in pressure was also established: "No employee shall be permitted to work in any compartment, caisson, tunnel or place where the pressure shall exceed fifty psi."[25] The century-old philosophies were finally changing. Workers were becoming more of an asset to be protected and cultivated rather than as a liability.

Between 1914 and 1921, the East River was again tunneled, this time by the New York Transit Commission, which employed Edward Levy (1878–1937), M.D., to direct the use of compressed air. The New York Transit Commission built several rail and gas tunnels between Brooklyn, Queens, and Manhattan between 1914 and 1921. All required compressed air. With many healthy men off fighting in World War I, tunnel-diggers and "sandhogs" in the earlier years (1916–1917) were often not ideally healthy, thus adding to the physical risks of the work. Levy, in contrast to Haldane and almost all previous investigators, believed that obesity was *not* a factor in decompression sickness, that temperature and humidity had little if anything to do with the bends, and that the Keays-Japp tables were too restrictive for taking advantage of a shift's work period. Levy, a graduate of Columbia University's College of Physicians and Surgeons, after years of community work became medical officer to the New York Transit Commission and then the Public Service Commission between 1914 and 1919. His paper on his observations during this work was published by the United States Bureau of Mines, which named him a consulting physician.

Levy used a staged decompression for pressures up to 45 psi, 5 psi more than the Pennsylvania Railroad tunnels, but he lengthened the work-shift time while decreasing the pressure. The incidence of the bends using Levy's tables was comparable to that of the Keays-Japp tables.[26] Levy reported that of 621,342 decompressions there were only sixteen serious cases.[27] However, he decompressed all of his workers to one-half *gauge* pressure, not absolute, and so he was probably a bit more conservative than he thought. Levy's medical jurisdiction covered the two tunnels built to 14th Street and 60th Street on the east side of Manhattan, where pressures reached 45 and 48 psig, respectively. Levy was, however, confident early. "Up to the present time [1916–1917], there have been more than 700,000 decompressions, and it is believed, from the results thus far obtained, that it is possible to work with safety under a pressure in pounds considerably in excess of the limit now fixed under the New York State labor laws, providing the working shifts and time and method of decompression are properly adjusted."[28] Levy's defiance may have been due, in part, to his working assumptions that there was no limit to the pressure, per se, to which one could be exposed: 50, 60, or even 70 psig. The limitations would be on the partial pressures of oxygen, nitrogen, and waste gases, as well as the sophisticated decompression schedules required for such pressures. None of these factors would be well studied for another half a century, and so early laws, engineers, unions, and doctors all struggled to make sense of the hazards of an industrial agent that no one could even see.

The public officers for New York's Labor Commission were first warned of the dangers of compressed air by the potential calamity of the 1892 Ravenswood gas tunnel under the East River, where air pressure once reached an intolerable 55 psig. Due to the tremendous variability in the decompression schedules of Keays and Japp and Levy and the many European tables used throughout the state of New York, and the unchecked dangers of compressed air and its purity, the state of New York finally stepped in again to make the system uniform. The American Association for Labor Legislation had officially adopted the modified Keays-Japp tables for uniform decompressions by 1914 which New Jersey passed as law. There were still great concerns regarding the dangerous nitrogen levels to which workers would be exposed with longer shifts. The medical professionals also had a strong desire to standardize requirements for workers in terms of physical health, age, weight, and medical problems.

Between 1918 and 1920, hearings were held in Buffalo, Rochester, Syracuse, Albany, and New York City, organized by the Industrial Commissioner, the Honorable Henry D. Sayer, and included Dr. Levy of the Public Service Commission, a number of insurance companies, representatives of the compressed air workers' union, compressed air work contractors, and several other physicians. As Commissioner Sayer himself believed, "[Former rules] contained many requirements which were applicable to work in compressed air and work in free air. . . . This made it necessary a complete revision . . . by striking out all rules which were in conflict."[29] A decompression table was established by Rule 1151 of the New York State regulations in 1921. The matter of staged decompression was legally stated in the Industrial Codes of the Department of Labor as follows: "DECOMPRESSION. No person employed in compressed air shall be permitted to pass from the place in which the work is being done to normal air, except after decompression in the intermediate lock as follows: A stage decompression shall be used in which a drop of one-half of the maximum gage pressure shall be at the rate of 5 lb. per minute. The remaining decompression shall be at a uniform rate and the total time of decompression shall equal the time specified for the original maximum pressure."[30] Although no precautions were made for the *length* of exposure at pressure, the improvements in worker health were quite impressive.

Adopted from the best aspects of Levy's tables and Japp's earlier figures, the regulations were used before they were actually made into law in the digging of the East River tunnels between 1914 and 1921 by the New York Transit Commission. Pressures reached as high as 48 psig and a total of 1,361,461 decompressions were required. The ratio of cases of decompression sickness to decompressions dropped from 1 in 346 for the Pennsylvania Railroad Hudson tunnels to 1 in 2,002 for the Transit Commission East River tunnels. Deaths dropped from 1 in 190,000 decompression to 1 in 680,730, respectively.[31] By 1922, the chance of incurring illness was one-sixth to one-thirteenth what it was in 1908, and the chance of death dropped to one-quarter to one-twenty-fourth of that found for the Pennsylvania Railroad tunnels.[32] The New York State tables were especially good for higher pressures. Between 1904 and 1909, there were *twenty times* as many cases at pressures up to 40 psig as there were up to 45 psig between 1914 and 1921.

New York's law of 1922 was eventually updated and adopted by several other states: Massachusetts in 1930, Maine in 1931, and California in 1933.[33] With the passage of the Wisconsin tables in the early 1930s, a worker in New York who moved to Milwaukee would find himself following different shift lengths, break lengths, decompression times, and maximum pressure exposures. It seemed as if each state was adopting slightly different and therefore confusing decompression tables. Unlike the U.S. Navy, which used one table of decompression rates for its hundreds of divers from San Francisco to Sri Lanka, the American engineering community seemed headed for having many "official" tables, each passed by a different state. Still, the New York State tables were the first tables in the world that were uniformly established, distributed, and adopted for use among the engineering community by a government or legislative body. These tables, representing the combined efforts of the medicine, labor, and engineering, contributed to the establishment of modern industrial health codes (fig. 35).

During the following years, long after a caisson worker's days

Figure 35. "Sandhogs" or work gang, Weehawken, New Jersey, during process of decompression in a chamber of the Lincoln Tunnel in 1934. Although some of the first laws regarding decompression had gone into effect, many of these "sandhogs" from the early twentieth century would be plagued years later by bone diseases caused by inadequate decompression. (Courtesy, Port Authority of New York and New Jersey)

Figure 36. The caisson of the Holland Tunnel ventilation building. Although designed in 1929, this caisson was as sophisticated as any used in modern tunnel or bridge building. Its huge compressors supplied over thirty million cubic feet of air per hour. (Courtesy, Port Authority of New York and New Jersey)

were over, it would be seen how grossly inadequate these tables still were, as they did not prevent long-term diseases like osteonecrosis (breakdown of the bone in pressurized air). However, by the 1920s, treatment for the bends was becoming an art. B. H. M. Hewett and H. Johannesson, engineers from the days of Greathead's London tunnels, thought that it was to the "advantage of the engineering profession, to say nothing of the ordinary motives of humanity, to see that the cases of compressed air illness are reduced to the minimum." [34] Hoping to put even more medical care in the tunnels, they wanted to "give the physician in charge a large degree of responsibility and authority and his word should

Table 1. Decompressions in New York Tunnels, 1902–1927

| Tunnel | Year | psig* | decom-pressions | cases | deaths |
|--------|------|-------|-----------------|-------|--------|
| Hudson | 1903–1907 | 42.0 | ? | 1,573 | 3 |
| Battery | 1902–1908 | 42.0 | ? | 674 | 14 |
| East River | 1904–1909 | 42.0 | 557,000 | 3,692 | 20 |
| 60th Street | 1917–1921 | 39.5 | 211,960 | 270 | 0 |
| Holland | 1920–1927 | 47.0 | 756,565 | 528 | 0 |

*Source:* Signstad, "Industrial Operations," 522.

*maximum pressure

be absolute in his department. The engineer's interest . . . is to see that the best qualified physician is retained and that he is encouraged to investigate and record the cases in the fullest degree. It is worse than useless to have carefully drawn rules treated as a dead letter. Make the men understand that the rules are there for their benefit and to save them suffering and death."[35]

Clifford M. Holland's great monster-caissons were sunk into the Hudson in 1920 to begin the Hudson River Vehicular Tunnel, better known as the Holland Tunnel (fig. 36). Like John Roebling's Brooklyn Bridge five decades before, the tunnel took Holland's life and tested the physical endurance of three subsequent engineers. It was completed by 1927 and carried thirteen million cars by 1931.[36] Edward Levy, by then the director of the medical division of the Port Authority of New York, oversaw the care of the "sandhogs." With sophisticated gas analysis of the carbon monoxide and car exhaust levels within the tunnel, he helped design the enormous air duct system in place today, which supplies over thirty million cubic feet of air per hour. His redesign of the Moir recompression chambers were thought to be the most advanced in the world for comfort, accessibility, and medical control.[37]

After months of *average* pressures of 30 psig, 2,243,546 decompressions and four- and eight-hour split-shifts around the clock, the 8,463 foot Holland tunnel reached New Jersey. The workers underwent decompression according to the 1921 tables, yielding fewer serious bends cases than the Transit tunnel of 1914 and no deaths at all.[38] It seemed, for a moment, that decompression sickness was finally beaten. Compressed air does not cause only the bends, however, and can cause diseases long after, even years after, exposure to it. Such effects of compressed air would tragically show how inadequate even the New York tables were, a deficiency that still stands largely uncorrected today.

Since 1888, it had been known that the bones of the body were particularly sensitive to decompression sickness. Haldane showed that bone and ligament are among the tissues with the slowest half-times and are particularly predisposed to bends injuries. By 1912, very few physicians made a distinction between the bone pain at the time of an accident and a crippling bone and joint deformity that could plague the same worker for years and years. Elrich Plate first identified bony lesions in bends victims with X-rays in 1910, months after the patients recovered from an attack.[39] Two years later, American neurologist Peter Bassoe may have been the first to suspect that either compressed air itself, or a bends attack of the bones, predisposed the tissue to some permanent injury.[40] He based this notion on his observation that of 145 men with the bends, eleven developed chronic arthritis and joint destruction several years after the original bends attacks.

Some of these were particularly gruesome. A sixty-nine-year-old carpenter worked in the Four Mile Tunnel in Chicago in 1879 and worked two eight-hour shifts at 30 psig in another tunnel in 1899. He decompressed rapidly and on the walk home the bends struck: a paralysis and pain of the legs from which he did not recover fully for seventy-two hours. He walked with a limp, permanently weakened ever after. By 1900, he needed a leg brace. By 1910, he had severe arthritis of the hips. The man was first

X-rayed by Bassoe's associate Hollis Potter, M.D., and the doctors found "all changes [similar] to arthritis deformans, namely: The head of the femur is flattened and mushroomed; cartilaginous space decreased and [bone spurs] laid down at the edges."[41] Another man who suffered from the bends worked as a bricklayer in the Rockford, Illinois, tunnel of 1897 when he decompressed from 39 psig in five minutes after two "long" shifts. He immediately lost control of his legs and could not walk for nearly six months. During that time, he could not move without excruciating pain in his muscles, which felt as if they were "all tied in knots."[42] Never free of the pain thereafter, he was wracked by arthritis of the knees and hips which developed insidiously months later. He finally gave up bricklaying. These and other sad tales were a testament to the subtle dangers of compressed air even if one survived the initial bends attack. Bassoe and his radiologic colleagues were so alarmed with their findings that they submitted a formal report to the Illinois Commission on Occupational Diseases in 1911. They did not prescribe any changes in the times workers would decompress, apparently hoping that their report would initiate such improvement in the engineering field. It did not.

By World War II, American pathologists used the microscope to examine bone biopsies of patients who had suffered from the bends and chronic limb pain.[43] The centers of the long bones, the areas of very active metabolic activity and growth but limited blood supply, were dead or severely scarred. The outer bone was overly calcified. Many of the diseased joints had lost their smooth, soft elastic cartilage and instead had flattened, rough, and scarred bone interfaces. The heads of the humerus (bone of the upper arm) and the femur (upper leg) were most often affected. These findings were of only medical and pathologic interest at first. Few realized that the prevalence of these destructive bone and joint lesions would be so high among workers that the decompression tables used over the past fifty years would have to be completely reviewed and revised.

By the 1950s, hundreds of men across the United States were now complaining of various pains in the legs, months to years after leaving a compressed-air working environment, even if they had never gotten a full-blown case of the bends. X-rays of these workers' femurs and tibias revealed the kind of degenerative patterns described by Kahlstrom and Burton in the 1940s and which often progressed to incapacitate a patient permanently.[44] Bone cells receive their nutrients and oxygen through tiny vessels that perforate the hard layers of the bone itself and travel deep into the bone substance through tiny tunnels. The tunnels, seen as holes in skeletal bone and even fossilized dinosaur bones, carry blood vessels which consequently cannot expand like other vessels. This makes the blood supply to certain parts of long bones extremely tenuous. Enough athletic or traumatic injury to the top of the femur in the hip can impinge so severely on the single artery supplying it that the hip joint itself can break down and fail (as happened to football and baseball player Bo Jackson). Compressed air apparently can cause similar damage to "watershed" areas of bone, either by slowing normal blood flow during decompression or by the action of higher partial pressures of the gases flowing through such delicate tissues. The condition of bone breakdown due to pressurized air, known as *dysbaric osteonecrosis*, remains the single most injurious, potentially permanent, and hazardous aspect of compressed air work. By 1958, in the building of the Clyde Tunnel in Scotland, 19 percent of the workers suffered from osteonecrosis.[45]

The problems of the split shift were twofold: the break between the two shifts was too short to allow enough residual nitrogen gas from the first shift to escape, and *two* decompressions were required for every worker every day, each decompression having the usual dangers of the bends.[46] The French and the Germans were quick to realize the importance of lengthening the break time which, elongated to twelve hours, was not so much a "break" as it was a time of exclusion from the compressed-air environment.

Even a twelve-hour "break," according to some physiologists today, was still too short.

While the French and British plowed ahead, improved and studied the staged decompression theory, American labor unions accepted a longer, uniform decompression schedule written by G. J. Duffner in 1963. Eventually known as the Washington State tables, they became locally popular. Duffner had originally written the tables as a staged decompression which he believed had merits over uniform decompression.[47] By 1969, a group led by Eric Kindwall of the University of Wisconsin (trained at Yale Medical School and in the U.S. Navy) found that the rate of dysbaric osteonecrosis was becoming frighteningly high among compressed-air workers. While building a tunnel for a large sewer system in Milwaukee, 60 of 170 workers, or almost 35 percent, eventually developed bone necrosis, the highest rate at that time anywhere in the world of medical-engineering literature.[48] Seventy-one per cent of these cases were thought to be moderate to severe, and sixteen of the sixty men had such crippling disease that they were banned from further compressed-air work for life. These compressed-air workers, like almost all other American workers at the time (except for the Washington tunnelers), were still using a modification of the old New York State tables from the 1920s. As an emergency measure, Kindwall had the state of Wisconsin adopt the Washington State tables of 1963, hoping that it would reduce or eliminate osteonecrosis. With the apparent success of Duffner's tables, even the federal government began to take notice, and in the summer of 1971, they were formally adopted nationally

Now called the OSHA tables, they were received originally with enthusiasm, for finally there was a systematic, standardized approach to decompression from any pressure and any length of exposure. There was great hope that future workers would never hear of decompression sickness again. These tables, too, however, still had potentially dangerous flaws, especially for the mathematical "black box" of long exposures, times where even knowledge from the diving community is scant. Here, a tunneler could become completely saturated with nitrogen gas after working two back-to-back eight-hour shifts at pressures as "low" as 20 psig. The OSHA tables required a decompression time of 113 minutes, whereas Kindwall and others now recommend well over fourteen *hours* for a worker to safely "surface."[49]

By 1975, as they suspected would eventually happen, Kindwall and his coworkers found seven men in Milwaukee who came down with osteonecrosis after working at maximum pressures of 43 psig in a tunneling project, despite using the OSHA tables.[50] The rate of bone disease with the OSHA tables was therefore no better at these high pressures than the old tables from the 1920s. It was feared as well that bends cases as well as osteonecrosis were even higher, the true incidence falling victim to the caisson workers' desire to keep out of the doctor's office and on the payroll. Anonymous inquiries revealed that 5 percent of workers admitted to developing symptoms of the bends on any given work day at pressures between 19 and 31 psig, three times the reported incidence for those actually showing up for treatment, which was 1.4 percent.[51] The problem had become clear: no table existed that could help a bends victim for the higher pressures and longer working times. Finally, OSHA agreed to look into the problem, and it was Kindwall who spearheaded the effort to improve the system, bringing decompression table research out of the caisson and the tunnel and into the laboratory and the computer.

OSHA, through NIOSH, finally supported an objective study in 1980 by Kindwall and other physicians and scientists from the Undersea Medical Society and the Sea-Space Research Company to improve decompression tables so that even cases of osteonecrosis could be averted altogether (fig. 37). The realistic goals of these "new" tables were to eliminate serious bends cases and osteonecrosis, while providing safe decompression tables to 50 psig. Peter Edel of Louisiana had gained an enormous experience in is about 80 percent nitrogen). As Paul Bert once wrote, referring to the small gradient between body and atmospheric nitrogen, "nothing urges it out."[53] One method around this phenomenon, first practiced by German pilots in the late 1930s and then by the diving community, was to decrease the presence of nitrogen inspired by a patient by making him or her inspire pure oxygen. Here, the gradient between body and atmosphere should be maximized and decompression times, as shown by fliers and divers, greatly reduced. Oxygen not only increases the gradient for nitrogen release, it also helps red blood cells bend better through tight vessels and loosens the stiffened cell membranes of white blood cells after exposure to compressed air.

The U.S. Navy saw the wisdom of accepting oxygen decompression by the 1960s, thus reversing a previously unsupportive attitude prevalent since 1939 which cited fire and health hazards. A fire in a Japanese oxygen-repleted air lock in 1959 started by a discarded cigarette butt pointed to the dangers of oxygen decompression, which most contractors were not willing to accept. Still, oxygen greatly helped workers in the huge Queens Midtown Tunnel of New York in 1938–39, supported by landmark laboratory research by Dr. Albert Behnke, and was independently adopted by Germany, Brazil, and France in the early 1970s.[54] The United States and Great Britain officially stood by the air-based decompression system with all its own inherent dangers and inadequacies.

Sensing the importance of oxygen, Kindwall and Edel tested their Autodec III tables using low-pressure oxygen at maximum shift lengths and pressures up to 46 psig and found immediate improvements in their laboratory subjects. The eleven-hour decompression time of the Autodec III table and the four-hour OSHA time for a four-hour shift at 44 psig were reduced to just over three hours by Autodec III if the individual were using oxygen.[55] These were the first tables anywhere in the world then tested in the laboratory on healthy subjects *before* use on site. The

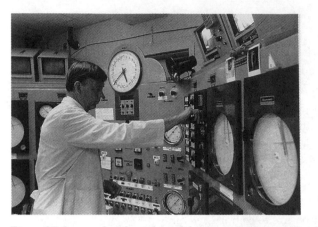

Figure 37. In a modern hyperbaric laboratory, Dr. Eric Kindwall
versity of Wisconsin works on his studies of decompression for use
Autodec III tables of the 1970s. (Courtesy, Eric Kindwall)

the world of commercial diving, which subjects indiv
much deeper pressures than any caisson worker ever had
not for as long a work period. Using a novel compute
Edel and Kindwall used fifteen years of diving data a
research to obtain a new decompression table which
longer, staged decompressions than the OSHA system. A
table, called Autodec II, failed in the laboratory under
trolled conditions, Autodec III was developed by Edel's c
It was theoretically and physiologically superior to tl
tables and promised to be the safest calculations yet. 1
decompression, however, were *very* long; the four-hou
pression time by the OSHA table for a four-hour shift a
was lengthened to almost eleven hours. As Kindwall adn
this were the ultimate price for tunneling . . . commer
pressed air work as we know it was doomed."[52]

The problem with nitrogen is that the only way it ca
pelled from the body is by lung ventilation which, in tu
ited by the large amount of nitrogen in our atmospher

laboratory data and physiologically based theories of the Edel-Kindwall experiments presented OSHA with irrefutable evidence of the superiority of oxygen decompression as early as 1983.

By 1988, Dr. Ralph Yodaiken, the director of OSHA's office of occupational medicine, admitted, "The current OSHA tables are inadequate for a number of reasons. They use a continuous decompression rather than staged, . . . based on [far fewer] tissue half-times . . . and use only air as the breathing medium during decompression." Reviewing the Edel-Kindwall Autodec III oxygen tables, Yodaiken believed that "there is no dispute within the scientific community that the oxygen tables would be a great improvement . . . and their use would lead to a much safer decompression. . . . Oxygen decompression is long overdue in caisson work . . . the current OSHA tables are out-dated and should be replaced in their entirety." [56] Despite these propitious conclusions, changing a system that had been indoctrinated into the union- and contract-based world of compressed-air work has not been easy. Currently, compressed-air workers, despite space-age robotics, high-tech shield designs, and on-site medical care, still expose their bodies to a decompression system established in 1963. The perceived dangers of oxygen by the labor community, the bureaucratic obstacles, and the realities of nationwide training in new decompression methods have made it very difficult to change decompression techniques substantially.

As early as the 1920s, the American and British navies pioneered tactics of decompression technology. With worldwide oil exploration booming after the second World War, commercial divers soon needed a state-of-the-art expertise in treatment of decompression sickness as well. Together, the methods of military diving units and commercial diving companies illustrate some of the most advanced theories and practices of compressed-air and inert gas use in modern times. The history of diving truly illustrates not only the past of compressed air use but the future of humankind's ultimate limits to their journeys on the earth.

# 12

## Diving Records and Rescues:

## Twentieth-Century Deep-Sea Diving

Early submariners realized that to venture into the ocean of water would require taking with them the ocean of air, or at least continuous access to it. Breath-holding meant short times in frigid waters, unprotected and with the crushing weight of deep water pushing against one's chest. An unknown thinker and beachcomber thousands of years ago first thought of applying the protective home of a nautilus shell for use in diving. These shells, known eventually as "diving bells," protected a diver descending in waters and also trapped air within it for the trip down into darkness and back up to the surface. As Aristotle observed well over two thousand years ago, open diving bells were limited by the compressibility of the air at the ceiling dome and with each foot of descent, the atmosphere of air was compressed to smaller and smaller volumes. Their relative safety relied on the weight of the water outside the bell being equal to the pressure of the atmosphere of the diving bell dome. A closed system would allow the entire weight of the water to be exerted on the walls of the bell itself while the bell atmosphere could escape compression. This was more risky, but with strong enough materials, early divers saw the merit of "one atmosphere diving."

would rarely occur in less than a hundred feet of water, many feared that even Haldanian tables would be inadequate.

The frustration of not easily being able to reach depths comfortably to save the sailors of the U.S.S. *109* fueled a heightened effort to break the barriers imposed on divers by nitrogen. Nitrogen is dangerous not only because it can cause the bends but at a high enough partial pressure can act as a kind of narcotic. In the 1940s it was found that nitrogen was the active component in all inhaled anesthetics, like nitrous oxide (also known as laughing gas). Gaseous molecular nitrogen is no different. At partial pressures above 8 atmospheres, nitrogen concentrations can have a serious if not crippling anesthetic affect on the brain. At depths of 180 feet or more, this effect, called nitrogen narcosis or "rapture of the deep," in which judgment, movement, and vocalization can be impaired, often leads to injury or death. This limits the use of compressed air to depths less than about two hundred feet. North Sea oil rig divers are, for this reason, not allowed by law to go deeper than 165 feet when breathing compressed air.[8] Even by today's standards, Frank Crilley's dive to 306 feet in 1915 is a kind of record for depth using compressed air.

With greater depth, not only must nitrogen be decreased but so must the volume of oxygen. The partial pressure of oxygen at sea level is about 160 mm Hg, and with each additional atmosphere the partial pressure increases proportionally. Oxygen, despite its vital importance to life, can be quite toxic as well. Even at sea level (where the partial pressure of oxygen is 160 millimeters of mercury [mm Hg]), breathing greater than 60 percent oxygen for more than six to twelve hours risks lung damage. At higher partial pressures, the nervous system is affected. Oxygen-breathing divers know well that twinges or twitches, though seemingly harmless, are an early sign of oxygen toxicity. At about 8 atmospheres (partial pressure 1,300 mm Hg), oxygen can cause convulsions, seizures, and death. Somehow Crilley at 306 feet for about ten minutes escaped these symptoms as well.

To use oxygen in deep dives, therefore, the intended depth must be predetermined and the oxygen volume of inspired gases proportionally lessened so that at depth the calculated oxygen partial pressure ranges safely between 160 to about 450 mm Hg. To decrease the volume of nitrogen, ideally to zero, and to lower the volume of oxygen would therefore require another gas to fill the deficit, a gas with none of the toxic effects of oxygen, none of the narcotic effects of nitrogen—a gas which, in short, has no chemical or metabolic activity at all. Such a definition describes the elemental inert gases, or "noble" gases, named for their apparent "independence" from any other atomic structure. Although the narcotic effect of nitrogen was not demonstrated by diver-scientists until the 1930s (the first being Albert Behnke), many had already grappled with solving the problem of oxygen toxicity. A reserved, well-educated theorist named Elihu Thomson thought he had the solution to the oxygen problem but unknowingly paved the way to solving the nitrogen problem as well.

Thomson first contemplated noble gases for use in diving during his work with the United States Bureau of Mines as early as 1919. Thomson realized that helium, nature's second lightest gas, was always physiologically harmless and safe. Using helium as the volume occupying gas in diving, Thomson reasoned, would theoretically increase the diver's time on the bottom with a much broader margin of safety from the effects of high-pressure oxygen.[9] Other than highly reactive hydrogen, helium is the most plentiful element in the solar system. Prehistoric collections of helium had been locked in the earth's crust for billions of years and could be extracted from such uranium-bearing minerals like pitchblende with painstaking efforts. By 1900, a cubic foot of the chemically worthless stuff could fetch $2,500. Then, in 1903, a subterranean collection of the gas was found near Dexter, Kansas. Several years later, the largest mine ever discovered was found in Amarillo, Texas. This single mine, under the jurisdiction of the United States Bureau of Mines, is the world's main supply of helium. Elihu Thomson originally had considered the diving problems of oxygen in the 1910s but did not write to his friend J. C. McLennan about using helium to replace air until 1919.[10] This insight, which few shared initially, eventually won worldwide acceptance.

With the need to go deeper, the U.S. Navy reconsidered Thomson's proposal and in 1924 the Navy's Bureau of Construction joined with the Bureau of Mines in Pittsburgh, led by R. R. Sayers, W. P. Yant, and J. H. Hildebrand, to begin preliminary research on helium-oxygen mixtures in animals.[11] By 1927, however, results were quite discouraging. Admiral Monsen, inventor of a submarine escape breathing apparatus (the Monsen Lung), noted that divers using helium at high pressures for long dives could still get decompression sickness.

Still, the Navy considered helium-oxygen an important mixture, but put emphasis on other work because of these early frustrations with it. A new Experimental Diving Unit (EDU) in the Navy Yard in Washington, D.C., was joined the same year by the U.S. Navy School of Deep Sea Divers. The Milwaukee physician and diver Edgar End felt somehow so convinced of helium's importance that his own research disregarded the official stance of the EDU. End found that helium, unlike nitrogen, diffuses very fast in the body's tissues. To avoid the bends, divers would need to allow helium to escape at depths far deeper than that calculated for nitrogen. These "deep stops" figured so prominently in End's research that he realized the only diver qualified to test such results would be himself. He found that his decompression tables using helium was so safe that he eventually trusted them on someone else. In 1937, End directed the young MIT-trained engineer Max Gene Nohl in a world record dive to 420 feet in the dark, cold waters of Lake Michigan. Nohl, alone in the abyss, followed the decompression stops as Dr. End had calculated, each with life-preserving importance. Eventually, the young Nohl surfaced without any untoward effects, having set a world record for that time for a dive in open water.[12]

Buoyed by such success, the U.S. Navy and the EDU reconsidered helium-oxygen using End's reconfigured tables. By 1939, helium-oxygen, or heliox, diving had become well-tested. With the discovery by Albert Behnke of the narcotic effects of nitrogen in 1934, heliox was found to be a solution to the debilitating effects of pressurized nitrogen as well. Progress continued with naval research in the United States so that from 1939 to the late 1950s, no other nation had any experience with heliox diving. Academic physiologists from American universities later saw helium-based diving as quintessential physics applied to the human system.

The Navy's helium research finally paid its big dividend when the U.S.S. *Squalus* sank on May 23, 1939, with its main induction valve open in 243 feet of sea off the Isle of Shoals off the coast of New Hampshire. Steaming quickly northward, carrying a newly designed McCann submarine escape chamber, the U.S.S. *Falcon* arrived within twenty-four hours. The rescue chamber reached the *Squalus*, and a diver, using a combination of helium and oxygen, secured a downhaul cable to the torpedo hatch. Thirty-three sailors were thus rescued and brought back to the surface, sunlight, and safety.[13] The stern compartments of the *Squalus* were then accessed, but the bell rescue party was stunned to find that whole section of the ship flooded and all of the sailors there drowned. Moving on solemnly with their work, divers began salvage operations, and immediately found that at 243 feet using heliox, there were no impairments of sensation or decision-making. Work proceeded so well that by September, the *Squalus* had been floated and towed to Portsmouth, where it was refitted and rechristened, and as the U.S.S. *Sailfish* it sailed to a distinguished war record in the Pacific during World War II. The lone diver that originally had secured the first cable to the torpedo hatch was later awarded a Congressional Medal of Honor for his heroic efforts.

By 1945, Jack Browne (a Milwaukeean like Edgar End) reached 550 feet for a new world record in a diving tank.[14] With the impetus provided by the Navy, international and nonmilitary uses of helium in diving increased with feats of equally impressive bravery and reliance on one's own calculations. Divers from the Royal Navy finally made Great Britain's first heliox dive to 600 feet by the close of the 1950s, and by 1962 American Dan Wilson dove to 420 feet on heliox. Sent down by a commercial firm off the Santa Barbara Channel in California, Wilson's dive inaugurated the presence of the commercial industry in heliox use. By 1965, commercial diving, in many respects, had superceded the technological advances of the U.S. Navy.

The United States Experimental Diving Unit through the 1950s became the focus of all compressed-air, helium, and diving research for the U.S. Navy. Unlike their tunnel building relatives, the military and then the commercial diving community put much effort in advancing compressed-air technology. This disparity in research stems partly from the fact that many open-sea dives simply go deeper than any caisson worker would ever need venture. Such depths and pressures underwater not only make decompression dangerous but the compressed air state as well. The development of heliox during the late 1930s obviated the occurrence of nitrogen narcosis so that the diver's time at the sea bottom would be safer and more productive. One still had to contend, however, with the trip back up to the surface. In addition, where oil field divers and high-tech navies became familiar with helium, most of the rest of the world, as is the case today, still used regular compressed air. So despite the apparent adequacy of Haldanian staged decompression for most dives, cases of the bends and full-blown decompression sickness still occurred frequently enough that the Experimental Diving Unit studied how best to treat these patients.

In 1945, new therapeutic tables released by the Navy Experimental Diving Unit for treatment of the bends were thought to be at least nine times better than previous treatment schemes and were the standard throughout the diving world until the 1960s. A stricken diver after surfacing would be taken into a medical lock and recompressed back to a "depth" of 100 to 165 feet for an hour or two and then decompressed on a standard treatment table from that depth. Known as U.S. Navy tables 1–4, they unfortunately required that a bends victim be recompressed with air. This meant that 80 percent of the inspired gas used during the actual decompression was still nitrogen, the same molecule that the decompression treatments were supposed to safely remove from a patient.[15] Although air-based decompression was not the best method to safely achieve *maximum* rates of nitrogen removal, only 6 percent of bends cases did not respond to treatments used in 1946.

While the EDU was fast at work perfecting hard-hat or surface-supply diving across the ocean, scientists and engineers were bringing to fruition the dream of freeing the underwater diver from surface-supplied air altogether. Jacques-Yves Cousteau (1910–1997) was raised in the coastal town of Saint-André-de-Cubzac where proximity to the sea instilled in him a life-long devotion to marine study and exploration. Originally joining French naval forces during the Second World War, Cousteau and his colleagues appreciated the military implications of non-surface-supplied, *independent* swimming. Cousteau himself likely envisioned such technology as the doorway to the beauties of the submarine world. Cousteau worked on developing a reliable regulator which could sense the inspiratory efforts of the diver before any subjective strain could be developed and thus allow the passage of compressed air to proceed into the diver's airways. The regulator would then shut, allowing the diver to exhale directly into the water. With the mechanical intuition of his friend Emil Gagnan, the two scientists produced their first prototype regulator in 1943.[16] The design of Cousteau and Gagnan eventually became today's self-contained breathing apparatus, or "scuba." With gradual improvements in the air-regulator system through the early 1950s, scuba transformed the world of diving. Although the better insulation of surface-supply (hard-hat) diving allowed for longer work periods, scuba provided freedom. With their new-found mobility, scuba divers enjoyed a submarine independence only dreamed of by adventurers for thousands of years. The beauty of undisturbed reefs, wrecks, and caverns, previously inaccessible to air-hose-restricted surface-supply divers, were now the scuba divers' playground. In addition, with the relative simplicity of most air-based decompression schedules, divers with even a rudimentary understanding of math and physiology could safely dive. Scuba, therefore, opened up to leisure divers a world that had been exclusively owned by the Navy and commercial firms.

Soon, the deep was visited by vacationers, explorers, adventurers, or anybody with the unexplainable yet persistent desire to venture back into the seas from which we all evolved. With the explosion of leisure compressed air use, however, sloppy divers failed to follow the model of care and precision set by naval divers. Multiple dives, carelessly planned and prolonged bottom-times, and shortened ascent times became commonplace. By the early 1960s, bends and decompression sickness cases skyrocketed.[17] Unlike naval accidents, in which cases a recompression chamber was often available directly on ship or in port, leisure civilian accidents often occurred miles from any medical facilities, let alone those with recompression technology. Consequently, when patients were able to get medical attention after a typically long delay and were treated with the standard U.S. Navy treatment tables, they frequently failed to get relief. Failures to cure were becoming epidemic. The 6 percent failure rate of recompression for serious bends cases using tables 1–4, as seen during 1946, had by 1964 climbed to 47 percent.[18]

Two tables were reserved for serious cases (neurologic symptoms, paralysis, persistent and severe pain, etc.) and for those unfortunate victims of gas bubbles actually blocking the circulation to the brain. Having escaped from small tears in the lung lining, such air bubbles cause a rapidly fatal condition known as air embolism and is one of the most dreaded aspects of decompression sickness. Air embolism is a great risk for those who ascend too quickly, especially sailors struggling to escape from sunken submarines. In fright, such victims close their glottis and allow lung expansion to occur unchecked as they rise in the water. Air embolism occurs because the lungs tear or rupture; this can occur with as little as 80 mm Hg pressure (about three and a half feet of sea water pressure drop) across the lung lining. With images of suffering divers unable to move their legs and spending up to thirty-eight hours and sometimes days without relief in the claustrophobic confines of a recompression chamber, the research community went back to work.

Scientists revisited Albert Behnke's research using oxygen in the late 1930s, which allowed the first use of oxygen for more rapid decompression in tunnel air-locks in 1939.[19] Behnke and his assistant A. L. Shaw first recompressed bends victims to 165 feet (6 atmospheres absolute) breathing compressed air. He then returned the patient to 2.8 atmospheres but this time allowed the patient to breathe pure oxygen via a mask. With zero nitrogen in the inspired gas, the clearance of all the absorbed nitrogen was rapid and efficient. Cure rates were exceedingly high, and this was reflected in the great success of the Queens-Midtown tunnel builders for staving off serious bends cases.

Oxygen can be toxic as well as quite hazardous in terms of fire. Although not flammable itself per se, gaseous oxygen is important in nature for its support of combustion. (The 100-percent oxygen environment of the Apollo 1 capsule allowed even the smallest of sparks to grow into a raging inferno that claimed the lives of three astronauts in 1967.) Owing to these concerns, the U.S. Navy avoided using oxygen in decompression until the failures of the 1960s became evident. M. W. Goodman and R. D. Workman finally developed new treatment strategies called tables 5 and 6. Table 5 was for patients who suffered from bends pain only, known as decompression sickness type I, which includes joint or muscle pain and does not threaten the person's life.[20] Type I patients were recompressed at sixty feet and, if the pain went away in ten minutes, they were then given oxygen to breathe at 2.8 atmospheres absolute (ATA) for an additional thirty minutes. Table 6 was for patients who did not get better on table 5 and for more serious cases known as type II, which includes any type of apparent nerve injury, paralysis, dizziness, or respiratory difficulties. DCS II patients are clearly much sicker and are at the highest risk of death. Patients on table 6 were taken to sixty feet, where they breathed oxygen for one hour and were then brought to 1.9 ATA for a total of four hours of oxygen.

Research in the late 1960s by Charles Waite of the U.S. Navy revealed that even air embolism victims could be cured if patients were suddenly recompressed or "bounced" to as deep as 165 feet for a short stay followed by oxygen breathing.[21] To train sailors how to escape from submarines, giant diving towers became commonplace at Navy diving schools. The tower at the naval base in New London, Connecticut, could be seen for miles until its demolition in the late 1980s. In these towers filled with water, some over one hundred feet tall, divers would practice "escaping" from the bottom where they would enter the chamber through special air-locks. Some unfortunate sailors, not allowing the air to escape from their lungs as they ascended, would sustain an air embolism. Waite's recompression strategy of immediate "submersion" in the air-chamber could get these sailors cured and safely out of the treatment chamber. The U.S. Navy was greatly impressed by the cure rates and theoretical promise obtained by these researchers as they allowed even a 38-hour decompression for a serious DCS II case on compressed air to be shortened to as short as 6.5 hours on oxygen. The Navy and then the world finally accepted the Goodman-Workman tables 5 and 6 and Waite's air-embolism tables (known as 5A and 6A) for military use by the summer of 1967.[22] The problem with oxygen therapy is that pressures higher than 3 ATA cannot be used because lung damage and grand mal seizures can occur. With bubble formation in the blood, there seemed to be still no good balance between high pressure to lessen the bubble load and tolerable oxygen pressure to maximize nitrogen escape during decompression.

Along with civilian leisure divers who discovered the treasures of the tropical submarine world, commercial divers discovered the treasures of submarine petroleum and natural gas reserves. At the sea bottom, commercial divers would soon become the most credible source for long exposure dives and decompressions from the point of complete body saturation with nitrogen or more commonly helium. They knew well the drawbacks of the previous seven tables for decompression sickness treatments.

Xavier Fructus finally tackled the low pressure oxygen-high pressure treatment problem while at work for the French commercial diving company Comex. Fructus attempted to raise the partial pressure of oxygen just enough but combined with nitrogen as well. On Fructus's table, victims would be recompressed first to one hundred feet on a 50-50 mixture of nitrogen and oxygen and then slowly decompressed over a seven-and-a-half-hour period to the surface. The tables, known as Comex 30, were highly popular and used by the oil drilling community of the North Atlantic and North Sea waters as early as 1973.

Despite attempts to reproduce real-life occurrences in a controlled setting, the laboratory is not reality. The ideal decompression sickness treatment table is difficult to develop because all divers have different circumstances surrounding their injury; delay to reach a treatment facility, uncontrolled depths and air pressures used, and patient failure to report early for simple cases. This variation made even the sophisticated approaches of the modern oxygen tables fail too often, some by more than 90 percent of the time by the early 1980s. A particular problem arose when Taiwanese fisherman divers developed both type I and type II cases and were not able to get to the appropriate medical facilities for treatment. Decompression researchers of the Taiwanese Navy were left to help these bends victims many hours and sometimes days after the first pains, paralyses, or pneumothoraxes (lung collapse) occurred. Physician-scientists H. C. Lee and K. C. Niu decided to use the strength of the Waite 6A oxygen table, but instead of the Comex 50:50 mixture used a 60:40 mixture of nitrogen and oxygen. The patient would then be "plunged" to a pressure of 165 feet. Instead of decompressing quickly to 60 feet, Lee and Niu took slow 30-foot stops over forty minutes, or ten times longer than Waite recommended. Almost 75 percent of the patients had arrived at the treatment facilities a full forty-eight hours or longer after the initial event (some arrived four *days* later). Each diver had a different exposure time and degree of illness. Despite these challenges, the Lee-Niu table reached cure rates as high as 71 percent for hundreds of victims.[23]

The sophisticated approaches to such difficult cases using computer-controlled and closely monitored mixtures of nitrogen and oxygen are not easily applied in the field by recompression teams in isolated commercial or leisure diving. The Association for Diving Contractors (ADC), the largest regulatory body governing commercial diving ventures, believed in the age-practiced medical axiom of "first, do no harm" in providing treatment for victims of decompression sickness. Given the complexities that came with the advancements in decompression treatments, the ADC also adhered to the less formal axiom, "keep it simple, stupid." In the struggle to crush bubbles into a harmless size, many in the ADC felt comfortable with higher pressure treatments for the bends since many thought that "bubbles have a memory as to the pressure at which they formed."[24] Others equally trained felt intuitively that low-pressure oxygen remained the strategy of choice, especially for long exposures at lower pressures.

Physicians treating the bends patients rely largely on symptoms and physical exam findings to ascertain how well a particular treatment in progressing. Given the subjective nature of such measurements, quantification of treatments is difficult. The pursuit and understanding of quantitative data to study a disease in objective ways has dramatically helped physicians better the means to treat the disease. Decompression sickness, caused by the sometimes transient evolution of sub-microscopic bubbles of gas in the tiniest of vessels, has eluded such quests. Attempts to identify nitrogen bubbles with ultrasound waves or to follow nitrogen washout rates in the lungs stemmed from the need to know how well a treatment was progressing, beyond just how a patient felt.

Despite the inherent difficulties of DCS treatments and the disparity among commercial and military decompression tactics, the majority of patients can be successfully treated today. Barophysicians can choose from an assortment of tables depending on the nature of a particular accident, disease severity, and physical condition (fig. 39). It is sometimes even appropriate to switch from one table to another during treatment to best take advantage of a patient's perceived degree of improvement. Kindwall suggests that straightforward bends cases be treated according to U.S. Navy table 6 (recompressed for 60 feet using oxygen) if it has been less than six hours after the accident. If this fails to cure the patient's symptoms, the physician can switch to Fructus's Comex 30 table and bring the victim to 100 feet on 50-50 helium-oxygen. If still no improvement in symptoms is achieved, even in as little as ten minutes, the physician can switch again to the Taiwanese table at 165 feet using 60:40 helium-oxygen.

Following ADC guidelines, divers with the bends would be seen by a physician who would determine only two things before treatment was started: Does the patient have simple or serious decompression sickness (DCS I or II) and when did the accident occur (during or after a dive or during a treatment)? For severe symptoms and an emergent need to start treatments, patients would be relegated to tables of higher and higher pressures. Simple cases would be treated on U.S. Navy table 6. For persistent cases, either the Comex 30 or Taiwanese table could be used. For the most se-

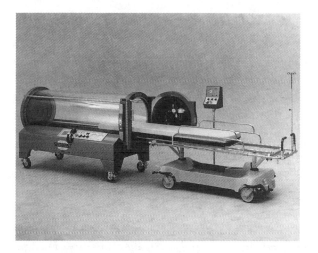

Figure 39. The result of a century of research and engineering, the modern decompression chamber is used to treat divers with the bends as well as a number of diseases in non-divers. Hyperbaric oxygen therapy is an important treatment for patients suffering from carbon monoxide poisoning, surgical healing, tissue damage from radiation, gangrene, bladder hemorrhage, and crush injuries. (Courtesy, Sechrist Industries)

vere cases, such as a total bailout (emergency ascent from depth), serious disease sustained after a very deep dive (greater than 165 feet) or worsening of symptoms during treatment, the ADC recommends even higher pressures for treatment according to tables 7 and 7A. These tables, or "saturation" ecosystem tables, became the ADC's secret weapon and require a sudden plunge to 200 feet or greater for thirty minutes and then a one-minute rise to 165 feet.[25] First developed by C. J. Lamertson in the 1970s, some diving medics think it is "amazingly effective and practical. One can enter it at any depth, on air up to 200 feet, on helium mixtures as deep as necessary. It is the solution . . . for all those 'nightmare' scenarios which have risen in the past. . . . [Table 7A] is the therapist's magic bullet."[26]

A diver at 220 feet of sea water, for example, becomes entangled in steel lines at the base of an oil rig in the stormy North Sea. After a struggle in which his coworkers cannot free him, his regulator lines suddenly become caught and tear. With only a few minutes more breathable gas left, his only hope is to "bail out" and get to the surface. Upon arrival on the ship, the diver complains of shortness of breath, agonizing knee and wrist pain, and his left leg has gone numb, the hallmarks of a serious accident (DCS II). Most would agree that treatment at 165 feet, even using high percentage oxygen, would fail to stave off catastrophe. Quickly he is put into the rig's on-site medical lock where the therapist begins treatment using table 7A. The diver is immediately "plunged" or repressurized to 250 feet. His symptoms improve and a small margin of safety has been claimed. Now, using a mixture of helium and oxygen, the diver is finally decompressed over several hours' time to 165 feet. He still feels well after a period of observation. Later, he is switched to table 6 with the "surface" within sight.

The medical, step-wise treatment of the bends from Ernest Moir's empiric attempts in the 1890s to the magic bullet of the Ecosystem 7A table of the 1990s has been a victory for the human journey into the subterranean and submarine worlds. The coordinated efforts of a tremendous number of scientists, physiologists, engineers, and divers themselves are to be credited with the evolution of modern diving tables. Compressed air breathing can be used safely by the majority of divers, from the vacationer to the high-tech oil line repair diver.

By the 1960s, diving physiology was studied by many academic scientists, as it illustrated both the biological commonality of marine and terranean life but also had immediate applicability. The ocean occupies more than 70 percent of the earth's surface. Below the surface of the sea lies 99 percent of the world's biodiversity.[27] To the scientist this represents a tremendous unknown in which one can study the variability of life, evolutionary laws, and the biological basis of life itself. To the world's navies, the ocean represented a huge battlefield where military dominance was a measure of how deep one could spy, defend, attack, or escape. To the commercial diver, the ocean bottom represents the forests of hundreds of millions of years ago, the graveyard of billions of tons of algae and plankton, all of whose remains have been transmogrified into the black gold of oil.

For these terrestrial mariners, the sea represents great promise. Overcoming the pressures of the deep remains their greatest challenge. In some areas of the ocean, a peak the size of Mount Everest can be sunk, its summit a mile below the surface. Each hundred feet exerts more than forty pounds per square inch. Compressed air could get a diver to 180 feet, maybe 200. Frank Crilley somehow reached 306 feet. Only 5 percent of the ocean bottom, however, is shallower than 600 feet. To reach the edge of the continental shelf at such depths, divers would somehow need to survive submersions nearly *twice* as deep as Crilley's. Clearly, compressed air would not allow divers to access the deep. By the 1970s, the search for the ideal gas was on.

Many institutions have sought to better our understanding of the deep. The Experimental Diving Unit (EDU) of the U.S. Navy and the Royal Naval Physiologic Laboratory at Alverstoke, England, were rivaled by few other laboratories for diving research. Continuing the on-site studies during the building of the Clyde tunnel in Tyne during the 1950s, the British National Research Council established the Decompression Sickness Central Registry at Newcastle University, which boasted the world's largest X-ray catalog of victims of the bends. In 1969, Duke University formerly established a facility now headed by Peter J. Bennett, which began a continuing tradition of pioneering efforts in diving gas physiology. In the mid-1960s, the State University of New York established a facility at Buffalo now headed by Claus Lundgren. Swiss researchers like Albert Buehlman explored high-altitude diving and set up high-tech facilities in Zurich while the French and Germans continued their commercial research in breaking the Envelope of the Deep. These investigators set out to study just how far down helium could take them, if indeed there was any limit at all. Using helium and oxygen mixtures (heliox), commercial divers broke one thousand feet by 1972.[28] Like Triger unlocking vast reserves of coal with his caissons in the 1840s, these workers began tapping reserves of oil which dwarf those previously found in the continental United States.

Despite the apparent "nobility" of helium, it too can eventually saturate the body and can be as almost problematic as nitrogen. Its benefit stems from the fact that it is less soluble in blood than nitrogen and clearly much less soluble in fat. At pressures of 15 atmospheres, where nitrogen would be clearly deadly, helium is well tolerated. Its liberation under decompression, however, can generate helium bubbles and can cause bends cases that are not dissimilar to nitrogen bends which occur after compressed air use one thousand feet closer to the surface. Decompression sickness from helium can occur, therefore, and for dives between five hundred and fifteen hundred feet, the bends represents the most dangerous aspect of helium use other than hypothermia from the dive itself. The French commercial divers for Comex had formulated their own ideal mixtures of heliox, but it was the EDU's work which developed the modern tables, known as table 8, for safe decompression from heliox use. These tables, published by 1993, represented the military and commercial standards, and they have been adopted throughout the 1990s.

More than 90 percent of the world's ocean floor lies deeper than four thousand feet. To reach such depths, there is clearly no way to test a human being's tolerance at greater and greater pressures. Apart from the ethical considerations, controlled experiments cannot be ideally performed on a ship in the middle of an ocean. Compression tanks or chambers were therefore built in academic research centers to permit controlled dives to deeper levels by simulating the effects of very high pressures (greater than 30 atmospheres or 21,000 mm Hg) which were unsafe to test in real water. These "dry dives" (and some in water-filled chambers as well) have allowed the world depth records for diving to be set although the "divers" were not actually in the ocean. Divers at Duke University led by Peter Bennett used their dry chambers to reach two thousand feet in 1972 breathing carefully controlled mixtures of helium and oxygen. Here, careful monitoring of vital signs, physical examinations, and physiologic measurements could all be carried out in the confines of a scientific laboratory. Should immediate decompression be required, no hazardous emergency "bail-outs" through freezing waters would be required. The divers in the dry tanks would simply need to "bump" up to decompression pressures, with immediate medical attention just a few feet away.

Bennett and his coinvestigators, however, soon reported that although decompression was not insurmountable given the correct calculations, the great pressures of the deep itself seemed to be causing a problem. Even a long exposure to compressed air at moderate pressures has no appreciable effect on a diver or tunnel worker. The scientists breathing very high pressure gases, however, experienced nausea, even vomiting, tremors, headache, prostration, and sometimes seizure-like activity.[29] These were clearly not evanescent complaints of the inexperienced, as they recurred on a reliable basis at pressures greater than six hundred feet when breathing heliox. This new phenomenon, called high-pressure nervous syndrome (HPNS), was found to be caused by a too quick compression to high pressure, usually greater than one hundred feet per minute to six hundred feet or deeper. Unlike nitrogen narcosis, which can often immobilize even the strongest, some divers can adapt over time to higher pressures and lessen the severity of HPNS. This habituation is severely inhibited at depths greater than thirteen hundred feet and it appeared, to Bennett and the Duke staff, that the two-thousand-foot limit seemed unbreachable. It appeared that very high pressure in and of itself was causing ill effects on the body.

Like serious compressed air bends, the symptoms of HPNS seemed to be neurologic: tremors, proprioception and balance difficulties, anxiety and nervousness, seizures, and catatonia. The main target of the syndrome seems to be nerve cells. Nerve cells have linings which are made up by a tremendous number of proteins. These proteins are rich in nitrogen molecules and nitrogen figures prominently in how these proteins help neurotransmitters and other chemicals relay information to and from the brain. These proteins, called channels, are arranged in important physical configurations on the cell surface and at cell to cell bridges. If protein channel configurations are deranged by a solvent, there can be deleterious consequences. Ether and alcohol are two of the most common solvents which affect neurotransmission, and in low doses their effect can be quite tolerable if not intentional. At higher doses, they can cause death. The fatty proteins of the nerve cell can apparently be affected by ambient atmospheric pressure as well. The greater the pressure, the more the protein arrangements can be distorted, particularly in the absence of gaseous nitrogen. The pressure's effects on the neural cell membrane and nitrogen-rich membrane proteins at the nerve synapse somehow influenced an important neurotransmitter called GABA (gamma-aminobutyric acid). GABA is thought to be important in keeping nerve cells from becoming overstimulated. High atmospheric pressure, as was used in the research chamber, seemed to inhibit the suppressive function of GABA molecules and this resulted in the neurologic symptoms of HPNS.[30] The exact mechanism of HPNS, however, is complex and will require more years of experience to pinpoint the exact defects in neural transmission.

Some researchers had noticed that mice poisoned by nitrogen could be reinvigorated with the application of high-pressure helium. Similarly, tadpoles intoxicated by ethanol recovered from the effect with the administration of 15 megapascals of pressure.[31] The reverse phenomenon, then, was thought to be true, that nitrogen could ameliorate the symptoms of high pressure helium. In other words, the symptoms of HPNS could be ameliorated by adding nitrogen *back* to the heliox. The resultant gas, called Tri-mix, enabled Peter Bennett to expose three divers to 2,250 feet of pressure in a chamber at Duke University in 1981.[32]

Helium promised deeper, long-term saturation dives than compressed air alone. Helium saturates the body almost three times as fast as nitrogen, however, so its use requires extremely judicious and well controlled decompression schedules. In addition, its lighter weight means it can absorb the heat from a diver's body more easily than the heavier nitrogen molecules of air. A certain amount of heat loss can be uncomfortable, and too much can risk hypothermia. With deeper depths, helium density increases so that more effort goes into respiring the atmosphere, risking inadequate elimination of carbon dioxide waste. An alternative was thought to be the only element lighter than helium: gaseous hydrogen. It is in long-term, low-oxygen partial pressure diving that hydrogen may be most useful. On hydrogen-oxygen mixtures, the work of breathing is lessened, heat loss is minimized, and with voice-decoders, even squeaky garbled speech in lighter-than-air environments can be understood.[33]

It had been shown in 1971 that HPNS in monkeys could be ameliorated by adding high-pressure, metabolically inert, hydrogen to heliox.[34] Although explosively flammable, hydrogen had been long considered for use in deep diving because of its light density (half that of even "lighter-than-air" helium), which makes it easier to breath even at *very* deep pressures. Its beneficial effect in HPNS was unclear. Complex mixtures of oxygen, helium, and hydrogen (hydreliox) are helpful in sophisticated research diving laboratories. Scientists at Comex in Marseilles, France, had safely exposed baboons to sixteen hundred feet (51 ATA) in 1972.[35] Humans reached only two hundred feet on hydreliox for several minutes by that year, but by the 1990s had gone deeper than the baboons. Several French researchers reached twenty-three hundred feet (70 ATA) breathing 45 percent helium, 54 percent hydrogen, and only 1 percent oxygen. Hydrogen, however, can become explosive when mixed with greater than 4 percent oxygen. For human diving purposes, hydrogen seems to be more dangerous than helium, and its routine use has not yet been widely accepted. Yet with the need to comfortably breathe, communicate, and work at very deep depths, hydreliox may have promise for the future.

Breathing *any* gas at 15 atmospheres absolute (ATA) of pressure increases the airway resistance, and the work capacity of divers can be considerably reduced. Some researchers like doctors Linnarsson and Fagreus of Sweden found that maximum aerobic power increased by 8 percent at even modest increases of pressure (1.4 ATA).[36] Others found that at higher pressure of 30 ATA the power level decreased by 13 percent.[37] One researcher, in a wet dive to fourteen hundred feet (43.4 ATA), decreased the maximum aerobic capacity by 22–30 percent, a severe reduction in the individual's ability to perform work for a length of time.[38] It is not known how deep or for how long divers will thus be able to go. However, these experiences at very high pressures make more realistic the dream of semi-permanent or prolonged submarine laboratory stays at depths as deep as the continental shelf.

Haldane's notion of tissue half-times was vital in developing the conception of the body as a complex mixture of different, heterogeneous substances. Diving physiologists as early as Haldane and Boycott knew well that exposure to compressed air for a long enough period would lead to complete nitrogen dissolution in every part of the body until no more can be absorbed. Decompression would be dependent on only the tissue with the slowest half time (bone and cartilage), a function of blood flow plus nitrogen or helium absorbed. The state of 100 percent absorption, or saturation, implied that once saturation was achieved, the decompression requirements from this total state would be the same no matter how *long* the compressed air exposure. That is, a diver saturated after a twenty-four-hour exposure at one hundred feet would require the same length of time (about 24 hours) to decompress to the surface as he would if he had stayed underwater five days longer. Obviously, the longer one stayed in the compressed-air environment, the more time one could spend at depth compared to the time trying to get back to the surface. Although a 24-hour decompression time is long, it is much shorter than the combined decompression times for twenty-four *separate* one-hour dives to one hundred feet, if that were possible. Such long-term stays, known as "saturation diving," truly promised an almost limitless stay at moderate depths, realizing the dreams of explorers who wished to venture into the submarine world.

Saturation diving was recognized early by Captain George Bond of the U.S. Navy in New London in 1957 as the key to make military, oceanographic, and commercial ventures into the sea more productive.[39] Bond, called "Papa Topside" by his Sea Lab Aquanauts, was the founder of saturation diving in the United States as Cousteau was for Europe. Originally a physician in his hometown of Bat Cave, North Carolina, Bond entered the Navy and quickly climbed in rank, becoming a specialist in submarine escape, "bailout" diving, and high-tech surface-supplied gas physiology.[40] Bond's work in the late 1950s in New London, Connecticut, with Walter Mazone and Robert Workman (who later developed treatment tables 5 and 6 for the Navy in 1967) confirmed that divers could live underwater on heliox for very long times. By 30 hours, with the body completely saturated with inert gases, the stay can be theoretically indefinite, limited only by the diver's other functions in life: nutrition, temperature regulation, solid waste, and, of course, boredom. As told to the great diving historian James Dugan, Bond envisioned submarine laboratories where a diver who is completely saturated could venture out from his pressurized domain to even deeper realms. Since no more inert gas could be absorbed, the diver could return to his pressurized laboratory without fear of decompression sickness. Bond shared a warm friendship with Cousteau, who hoped to establish his own undersea labs off the coast of France. During one visit to the Yale School of Medicine, Bond lectured on his work at the nearby U.S. Navy submarine base in New London. An interested medical student asked about the future of diving and Bond quickly sketched on a napkin his vision of these undersea habitats of saturation diving and experimentation. Bond went on to turn these sketches into reality.[41] The interested medical student was Eric Kindwall, who went on to became one of the country's most eminent diving medicine physicians.

The first submarine labs included Cousteau's Conshelf I, II, and III from 1962–63, the Man in Sea projects directed by the undersea explorer Edwin Link in 1964, and then the U.S. Navy's Sealab I, built in 1964–1965. In Sealab II (August 1966), saturated divers were exposed to 450 and 600 feet during one-hour excursion dives. Such depths reach the continental shelf itself; beyond that lies the great continental slope from 600 to 9,000 feet, where 95 percent of the undersea world lies deep in the dark coldness of the abyss.[42] For human advancement into the submarine inner space, the precipice of the shelf will somehow need to be left behind for humans to consider the mysteries of what lies below in the great depths. For such goals to be within reach, divers may have to abandon breathing gas mixtures altogether. Instead, humans may have to adapt the methods of respiration used by marine animals for eons: liquid breathing.

# 13

## Going to Extremes:

## The Ultra-Deep Ocean

*Homo sapiens* evolved in an atmosphere of 14.7 psi (1 ATA) over millions of years and have developed a means to exchange gases for aerobic respiration that are dependent upon a very narrow window of atmospheric pressure. The body of *Homo sapiens* similarly has a skeletal muscle and vascular system that is just strong enough to support its life on the gravitational field of earth. Other animals have a greater diversity for possible areas of habitation and this results from a much greater adaptability for environmental changes than possible in *Homo sapiens*. The Weddell seal (*Leptonychotes weddelli*), for example, is an air-breathing mammal which happily lives in the frigid waters of the Antarctic, feeding on codfish and other indigenous fish. But with a combination of enormous blood volumes (sixty liters/animal versus about six for an adult human male), blood reserves (storage by the seal spleen can account for more than half of the total blood cell mass versus an insignificant contribution in *Homo sapiens*), increased oxygen-saving myoglobin production, and blood flow regulation, these seals can dive as deep as five hundred meters and hold their breath for longer than an hour.[1] This points to an remarkable ability to tolerate pressures as great as 750 psi, which in humans would crush the chest as well as severely limit if not preclude effective cardiac output and brain blood flow. The seals, having evolved these systems over the past twenty million years, illustrate by comparison the great physical limits that the sea and space contain for human endeavors without artificial adaptations.

Researchers as early as the 1950s had realized that sea-dwelling fish, being mostly water, evolved in pressures to which they were equilibrated at every point of the body. Once gills evolved to the point that they could effectively exchange gases in water, pressure itself was generally not as important as food-chain dynamics, temperature, and oxygen levels.[2] The observation that water-laden structures were noncompressible led to the theory that an oxygen-containing fluid could be developed which would be "breathed" by a divers as they descended into the depths. With the hydrostatic pressure in the alveoli of the lungs equal to that of the surrounding sea, all of the pressure effects which preclude breath-hold diving would be obviated. The depth for diving breathing "liquid" could theoretically be dramatically improved, perhaps by hundreds if not a thousand feet deeper than had ever possible breathing compressed gases.

Liquid breathing was first tested in the laboratory by J. A. Kylstra and his colleagues, doctors Tissig and van de Madeni, in 1962 using rats.[3] A saline solution equilibrated with the rat's normal serum salt levels was superoxygenated and supplemented with a carbon-dioxide-consuming buffer. The rats were then completely submerged in the mixture. After a few gasps the rats started inhaling the fluid, and Kylstra found that not only did they survive but they survived after removal from the fluid. Although the waste gases were absorbed well, the oxygen levels were not adequate, and in 1966 Clark and Gollan repeated the experiments using a fluorinated hydrocarbon called FX-80.[4] FX-80 had twenty times the oxygen absorption of saline, yet its carbon dioxide absorption was not adequate. Still, Clark and Gollan took fourteen mice submerged entirely in and "breathing" FX-80 and pressurized them in a chamber to 1,124 feet. At such pressures, organisms breathing compressed air would surely need sophisticated decompression tables to allow a slow, meticulously monitored return to "sea level" without developing severe if not fatal cases of the bends. Liquid-breathing, unlike compressed gas breathing, does not rely on gases (which can be absorbed by the body) to help equilibrate the pressure gradient between lung and the surrounding sea. Instead, liquid-breathing prevents the squeezing of the lungs by high pressure because the liquid itself is noncompressible. Without any absorption of nitrogen, helium, or any other gas, decompression sickness does not occur. Clark and Gollan's mice could, they reasoned, be surfaced immediately. Not knowing what to expect, the two removed the mice from the pressurized chambers and drained the liquid from their lungs. All fourteen survived a decompression from a depth that none would have survived by breathing compressed gases. The same year, Kylstra decompressed liquid-breathing mice from thirty-three hundred feet to sea level in three seconds, a speed which he found equivalent to an ascent rate of 700 mph.[5]

These experiments proved the safety of liquid-breathing on the laboratory rat and mouse, but there remained the problem of optimizing the oxygen-carrying capacity of the liquid as well as maximizing carbon dioxide removal. A mixture of the fluorocarbon (which carried oxygen well) could be mixed with a buffered saline solution (which absorbed carbon dioxide well). Kylstra found that the water-immiscible FX-80 could serve as the "red blood cells" carrying oxygen into the lungs and buffered saline could serve as the "plasma" carrying carbon dioxide out.[6]

The resilience of laboratory rats and mice is unquestioned, as is the relative frailty of the human being. The respiratory lining of *Homo sapiens* is extremely reactive and will be injured by too much oxygen as well as by caustic agents like gaseous poisons, as in car exhaust. After exposure to an injurious agent, tiny lung sacs (alveoli) lose the ability to keep out fluid and remain inflated. The airways can rapidly become drowned in body secretions from the blood stream. The syndrome of life-threatening fluid release, poor oxygen absorption, stiff lung tissues, increased cardiac work, and shortness of breath is known as the adult respiratory distress syndrome. The syndrome can be precipitated by a variety of lung injuries, including the aspiration of fresh water during drowning, fat embolism, or direct airway injury by gases or fluid. Liquid breathing in human diving, therefore, may be limited by the ability of air-evolved lungs to tolerate the liquid mixture itself. The ideal liquid will have many requirements. It must deliver enough oxygen, absorb maximal carbon dioxide and be completely harmless to the delicate lung tissues.

From 1976 to 1990, liquid breathing was completely experimental, used primarily in laboratory rodents and dogs. The time had come for human application, but its first recipients would not be the brawny, iron-hearted divers for Comex or the adventurers in the abyss of the Pacific rim. The first human liquid breathers would be those who had already grown in and survived nine months of liquid breathing but could not adapt to breathing air: frail, premature infants. Some premature newborns have immature lungs which fail to keep the tiny airways and air sacs open enough for good air exchange. Amniotic fluid does not provide oxygen to the infant, this job being accomplished by the circulation through the umbilicus to the placenta. Amnion, however, is vital for the normal development of the lung lining. Devoid of "surfactant," the lungs remain stiff and water-logged, and death almost always occurs without intensive intervention, intubation, and breathing machines.

Some doctors felt that if these lungs tolerated liquid breathing for nine months, they might also tolerate liquid again after birth if the liquid could be used to promote gas exchange. The medical world looked closely at the laboratory research concerning fluorocarbon breathing. Thomas Shaffer's work on canine liquid breathing showed that lungs become more stretchable and elastic when exposed to the fluid. Shaffer's lab was also able to slowly perfect the fluorocarbon mixture and exchange rate so that adequate oxygen and carbon dioxide exchange would occur. The system worked well on dogs, but the fluid had never been used on humans. Several Philadelphia doctors found a small number of infants who had such severe lung disease that not even machine-assisted ventilation was successful. Without any other medical or surgical options, the children's only hope for survival was liquid.[7]

Using gravity to instill and remove the fluid via a catheter placed in the trachea, three children were oxygenated with perflubron, a bromide perfluoroctyl which carried 50 percent oxygen and 20 percent carbon dioxide. The children tolerated liquid-filled lungs well, and there was no evidence of any lung damage caused by the liquid. The children had severe, terminal cardiac and lung disease, and although all survived several days on liquid breathing none survived their diseases themselves. Still, such heroic attempts have fueled the hopes that liquid breathing is a useful and applicable technology . We may find that to fulfill our desire to venture into the sea, we may have to do without our ocean of air. We may have to devise our own liquid ocean to breath in our submarine journeys just as evolution had created for our life's journey during its first nine months.

The absolute limit to the depths may not be related to gas physiology at all. The cellular metabolism of land animals depends on an uncountable number of proteinaceous interactions. Many of these are shape-sensitive interactions, such as those between hormones and their receptors, nucleic acids and DNA-binding proteins, and enzyme subunits. Shape interactions are, in turn, affected by pressure.[8] Nowhere on the earth is the pressure effect on an organism better illustrated than in the ultra-deep ocean. There, in the cold, lightless waters as deep as 10,000 meters (6.2 miles) and at pressures of 900 ATA (13,300 psig), slow-moving creatures dwell, feeding off the detritus from livelier waters above or on the occasional morsel in the muddy bottom.

Carnivorous teleosts (bony fish) at 1,000 to 3,500 meters, like the angler fish, simply wait in the darkness, bobbing a bioluminescent lure in front of their open mouths, hoping to attract the next unwitting victim into its fierce jaws. Their lives are so focused on energy preservation that even reproduction is kept to a minimum. Male angler fish parasitize themselves on their female partners and degenerate into a pair of gonads which provide genetic material to the female when she ovulates. In five-degree Celsius water, these and other fish can routinely tolerate pressures ranging from 150 to 400 atmospheres or at least 5,000 psi. Such organisms truly represent the ultimate in evolution's adaptability to a given environment.

Abyssal teleosts and other deep-sea fish rarely have been brought to the surface alive, thus precluding significant studies of their physiology. Certain molecular statements about them can be made in comparison to surface dwellers. Cellular fat and protein molecules are adjusted in a cell's structure to maximize its vital membrane fluidity. With changes in pressure, this cell molecular makeup can be changed with tremendous effects on the moment-to-moment activities of the cell. Surface dwellers have cell membranes which become stiff at even moderate pressures (50 atmospheres or sixteen hundred feet). Surface-dweller nerve conduction and cellular ion fluxes are drastically inhibited when exposed to this pressure and this may, in part, explain on a molecular level, high-pressure nervous syndrome. Correspondingly, cells which evolved at high pressures may not tolerate lower pressures well, and this may explain why some deep-sea fish are never found in shallower waters despite similar food chains and temperatures.

The high pressures of the deep also tend to determine the way proteins congregate to form enzymes and how well a particular enzyme works. Without proper enzyme function, no meaningful chemical reaction in the cell can take place. Yet deep-sea fish enzymes all have a remarkable tolerance for an enormous range

Figure 40. As portrayed by Ed Harris in the 1989 Twentieth Century-Fox film *The Abyss*, the character Bud Brigman readies himself to plunge into the ultra-deep. If ever realized, liquid breathing would allow divers to descend hundreds if not thousands of feet into the depths. (Courtesy, Twentieth Century-Fox Film Corporation, 1997)

of pressures. Surface-dweller enzymes, however, have no such capacity. Even at two thousand feet, many enzyme catalysts dissociate into their nonfunctional protein subunits—pressures at which the deep-sea fish commonly feed in their shallowest ranges.

Also pressure-dependent are the proteinaceous tubes which drive every cell division or movement in the immune or skeletal system. In cells adapted to 1 atmosphere these tubes will fall apart at 2,000 meters and cell division and movement will stop.[9] How ultra-abyssal animals maintain their mitoses and cellular movements uninhibited in pressure up to 500 atmospheres (seventy-five hundred pounds, or more than three tons, per square inch) is completely unknown.

Human exploration of these depths was portrayed in the 1989 Twentieth Century-Fox film *The Abyss*, in which liquid breathing brought one fictional diver to depths approaching that found in the rim of the Mariana Trench (about thirty-five thousand feet) (fig. 40). Journeys to such depths may never be possible for surface dwellers even using liquid for breathing. Instead of simply preserving adequate gas exchange, one must also preserve basic cell function and metabolism, as the enormous pressures of the abyss literally squeeze the cellular life of the surface dwellers slowly to a halt.

# 14

## Air as Medicine

The use of compressed air as a therapeutic agent was attempted as soon as the gases could be bottled and sold in the 1840s. Finally abandoned as quackery by the medical establishment of the early 1900s, the use of hyperbaric pressure with oxygen breathing was revitalized eighty years later. Physicians had long cared for patients with gangrenous limbs, poorly healing surgical scars, and radiation burns using conventional treatments, but without much success. Deprived of normal blood vessels, such human tissues are starved of oxygen. Compression was a potential way to drive precious oxygen molecules into the depths of such tissues, which the patient's heart and circulatory system could not do. Hyperbaric oxygen and its use in modern medicine changed compressed air from a hazardous necessity of the nineteenth and twentieth centuries to a clinical therapy of the twenty-first.

The Great Plagues of Europe and the British Isles are now known to have been caused by Yersnia pestis infections disseminated by rats, fleas, and lice. To the seventeenth-century Parisian or Londoner, however, there was no visible attacker, no tangible evidence that anything other than the air itself was the source of the disease. Such contentions of disease were noted by Greek playwrights like Sophocles, who described the house of Oedipus as being overtaken by a *miasma*, or bad air.[1] It followed, then, that if poor air were the source of disease, then good air should be the cure for it. Certainly, enterprising contemporaries of Robert Boyle and Robert Hooke saw profit and fame in such beliefs. In 1664 a clergyman named Thomas Henshaw proposed a *domicilium*—an airtight room where the air pressure could be controlled with organ bellows and valves.[2] Patients would be treated for various maladies including fever (for which they would be decompressed) and chills (for which compression was prescribed). Henshaw does not appear to have tried out his concept, but the idea to use compressed air to treat mankind's suffering in atmospheric air was born.

Even the academic community realized the potential of gas therapies especially given Priestley and Lavoisier's discovery of oxygen and nitrous oxide. The English scientist Humphrey Davy was working on gas chemistry when he was asked to lead a "pneumatic institute" at Oxford in the early 1820s.[3] Although pressurized gas would not be used, the institute was the first to treat patients with a variety of disorders. Davy used nitrous oxide (laughing gas) for a variety of maladies with unsubstantiated success, but interest continued to find therapeutic applications for gas under pressure.[4]

The pressurized chambers of V. T. Junod in the 1830s were the first to actually subject patients to compressed air. By 1842, Junod had claimed success in treating a young girl with tuberculosis.[5] The copper chambers were claustrophobic, each a 1.4-meter-wide sphere that could hold one person or two. Junod's compressors, however, had no regulators, and with each piston stroke a wave of pressurized air to 1.5 ATA would be uncomfortably blasted into the chamber. Junod's treatments were, therefore, not as popular as those of his contemporary, the Frenchman Pravaz, who built chambers in his orthopedic clinic in Lyons around 1837.[6] With better valves, compressors, and mechanical knowledge, Pravaz's chambers were well tolerated and some, including himself, found great relief at 1.5 ATA from headaches and deafness.

Some of C. J. Triger's first workers in his caisson reported the relief from hearing loss as well as asthma and breathing difficulties.[7] Anecdotal reports of such improvements fueled the excitement for pneumatic therapy, and by the 1850s several well-funded pneumatic institutes were thriving. Physicians established chambers in Vienna, Stockholm, Baden-Baden, Berlin, Toronto, Rochester, New Haven, and Munich.[8] Joannis Milliet in the 1850s used 1.5 to 1.7 ATA for two hours followed by a one-hour decompression period to treat emphysema and a variety of infectious and noninfectious lung diseases.[9] Charging examination and admissions fees, Milliet could treat dozens of patients at once in his chambers which were equipped with sophisticated air-locks. After the thirty to forty treatment sessions Milliet recommended, it is unclear if it was Milliet or his patients who benefited most.

At the height of pneumatic therapy or "air bath" therapy in the 1870s, a veritable dictionary of maladies and afflictions were treated. The treatments, however, were not supported by scientific or comparative data. Some of these diseases included bronchitis, bronchiectasis and croup, restrictive lung disease from scoliosis, conjunctivitis and allergic eye reactions, emphysema and asthma, rheumatism and arthritis, hysteria, mania and hypochondriasis, amenorrhea, dysmenorrhea and all gynecologic bleeding disorders, whooping cough and influenza, anemia and chlorosis, heart muscle and valve disease, dyspepsia and stomach ulcers, snake bites, and even smallpox.

By far, most of the interest by the better trained physicians appeared to be directed to the treatment of tuberculosis. Consumption, or pulmonary tuberculosis and its skin form, scrofula, have been and continue to be sources of human misery and death. Junod's initial experience in treating a single case of TB in 1842 was not followed by any other reports of success. When the bacillus which caused TB was discovered by Robert Koch, there was great interest in pneumotherapy for the rampant lung disease. This treatment appears, however, never to have worked. Patients with TB in the late 1800s eventually flocked to the clean air of the sanitaria and the Swiss Alps where the air was actually at less pressure than the air down in Lyons or Paris. Still, C. T. Williams utilized pressurized air to 1.7 ATA in the respected Brompton Hospital for Consumption and Diseases of the Chest in London during the 1880s.[10] Williams and his colleagues felt that the blanching of the skin sometimes seen under pressure was a sign of the rerouting of the peripheral blood into the deeper organs of the body. Andrew Smith believed this was the secret to understanding the bends in Brooklyn Bridge workers.[11] Long disproven, now thought to be due to the effect of compression on the small veins of the skin, the rerouting theory at that time was quite popular.

By the 1890s, more than ten years after Paul Bert published his seminal work on pressurized air, the Western medical community, led by German physicians, began to apply the laboratory method to its scientific questions and provide a professional accountability for its practitioners. This self-scrutiny included reviewing the use of compressed air, especially for tuberculosis. It had become clear, with the landmark investigations by Koch and Waldenburg on the lung pathology of TB, that dangerous openings in the airtight lining of the damaged lung could be made with compressed air, causing a pneumothorax (lung collapse).[12] As the nineteenth century closed, more and more physicians, now trained in rigid medical school programs in pathology, anatomy, medicine and the new field of microbiology, lost their interest for pneumatic therapy (especially for tuberculosis), and the therapy was quietly dropped. Compressed air was briefly visited and on the grand, if not outlandish scale by Orville J. Cunningham. Cunningham, who was not a physician, designed and built a multistory spherical building with pressurized rooms for rent called the Steel Ball Hospital in Cleveland, Ohio, in 1928.[13] There, patients would undergo, among other applications of pressurized air, the "Cunningham tank treatment" to relieve heart ailments and chronic pain. Like its failed predecessors a century before, Cunningham's treatments had no medical basis, his customers stopped coming, and the business came to an end.

Throughout the rest of the world by the 1930s, however, the invisible therapy of air under pressure was replaced by the tangible and very visible medications and techniques developed throughout the twentieth century to combat disease. With the dramatic increase in diving and engineering innovations after 1900, compressed-air technology was used only to prevent and to treat the disease it invented itself: decompression sickness. With the sophistication and medical approach to compressed air use, physicians by the 1940s began to reconsider pneumatic therapy for other human diseases. For this reason, pneumatic therapy, more properly known as hyperbaric therapy, has often been called a therapy in search of a disease. The benefit of hyperbaric therapy in nondiving applications stems solely from the ability to increase the tissue pressure of oxygen. Only in diseases where local tissue supplies of oxygen are poor can hyperbaric therapy be considered to be of some benefit. Such conditions include carbon monoxide poisoning, where blood cells are so tightly bonded to highly avid carbon monoxide molecules that they become unable to carry oxygen. By the 1980s, hyperbaric therapy had returned to the clinical armamentarium, now using oxygen instead of compressed air.

Hyperbaric oxygen therapy uses pure oxygen at 2 to 3 ATA and can raise blood oxygen content ten times higher than normal. For this reason, Behnke and researchers first used the therapy to increase the removal of nitrogen from decompressing divers in the 1930s. Some patients develop air leaks in the lung due to mechanical ventilation or medical procedures, and if air bubbles find their way into the main outflow out of the heart, air embolism can occur, just as in divers during emergency ascents from the deep. Air emboli in the hospital, as in divers, can be treated with hyperbaric oxygen, and that is the treatment of choice for this condition when recognized early. Continuous patient monitoring and blood oxygen levels help prevent any of the deleterious side effects of high-pressure oxygen therapy.

Human tissues sometimes react to radiation therapy used to kill cancers with the formation of scars. Depending on where the radiation beam is focused, different tissues can be affected. Pain, bleeding, intestinal obstruction, fistulas, and infection are just some of the complications that attend the use of radiant energy. These reactive tissues are notoriously poor in blood supply as the irradiated blood vessels gradually scar and cut off the blood supply to at first small areas. The resultant low oxygen levels in the tissue begin a vicious cycle of poor tissue viability that is thought to be the cause of the misery of radiation injury. Radiated tissue is so poor in blood supply that even blood oxygen contents of 2,000 mm Hg fail to raise the tissue oxygen pressure significantly.[14] Hyperbaric oxygen therapy does, however, apparently induce new blood vessel formation in these damaged tissues and this helps wound healing tremendously.[15] In poorly healing radiated head and neck wounds and bladder inflammation from radiation exposure, the treatment has been shown to bring about healing in a majority of the cases.[16] For the plastic surgeon, hyperbaric oxygen therapy has been used at critical points of healing where increased oxygen tension can help turn a tenuous, at-risk area of skin or muscle flap into a viable, growing tissue on the way back to recovery. Gangrene, traumatic injuries, and some severe burns may soon be more frequent applications of the therapy, based on slowly accruing scientific studies.[17]

Since timing is crucial, patients must be near a medical center that has appropriate equipment and the trained staff to run the compression chambers. Often these medical centers, such as the ones at Duke University, Norwalk Hospital in Connecticut, and Milwaukee, are already used by the diving community. Most doctors can access one of the three hundred hospitals that offer therapy via helicopter from their hospital. Both Duke University and the Maryland Institute of Emergency Medical Services in Baltimore offer telephone-based networks to direct physicians to the closest unit. New Worldwide Web sites for each hospital that offers hyperbaric oxygen therapy appear every week, and the home page for the Undersea and Hyperbaric Medical Society and the Diver's Alert Network are other on-line resources.[18] The number of hospitals in the United States offering hyperbaric oxygen treatment is increasing by about twenty a year.

As long as humans can become sick and there are people to care for them, the earth's resources will be searched for the answer to their illness. Now, armed with scientific techniques, quantitative analysis and clinical trials, modern physicians continue to increase and focus those maladies treated with oxygen under pressure, thus making real the dream of Henshaw three centuries ago.

# 15

# High Altitudes and

# Decompression Sickness

J. S. Haldane's work was pioneering in bringing to fruition years of research to enhance the safety of diving to depth, whether it be in a pressurized sea of air or of water. The principal event, bubble formation, occurred because the level of supersaturation of nitrogen gas was higher than the ambient pressure to keep the gas dissolved in the tissues. This happens whether a diver goes from 10 atmospheres of pressure to the 1 atmosphere at sea level, or whether a pilot flies quickly from sea level to fifty thousand feet.

Dr. Yandell Henderson, the noted American physiologist and chairman of the Medical Research Board for the U.S. Army in World War I, was the first to raise the possibility of decompression sickness at high altitudes and described the probable similarity of "flier's bends" to caisson disease. The balloon flights of the 1860s and 1870s had reached heights upward of twenty-five thousand feet, but so gradually that there was negligible evolution of any dissolved nitrogen present in the balloonists' bodies. The fatal effects of high-altitude aviation in ballooning and the symptoms of mountain sickness were largely due instead to hypoxia (oxygen deficiency) and a metabolic alkalosis (base accumulation) due to hyperventilation. Henderson was a member of Haldane's now legendary Pike's Peak team, named for the mountain where early research in altitude sickness was performed. In his original work published in Aviation in 1917, Henderson wrote: "In order for bubbles to be formed it is essential, however, that the pressure with which tissues are in equilibrium should be lowered more than half its absolute amount in a few minutes," or the Haldanian rule of tissue half-times.[1] An aeromedical research laboratory in Long Island was established to research this phenomenon, but, given the experimental nature of flights to extreme altitudes by 1918–19, interest waned and in 1920 the laboratory was abandoned.[2]

Aside from human physiological limitations, the airplanes themselves had mechanical difficulties reaching high altitudes. Even by the end of the First World War, no aircraft could routinely fly to altitudes much higher than twenty-thousand feet due to poor lift characteristics, inefficient engine power transference, and aerodynamic designs. Air is extremely thin at such heights and to achieve proper combustion in air-hungry engines, powerful compressors are required to suck in much larger volumes of air than the airplane normally encounters by sheer forward motion. The first plane to be outfitted with such machinery was U.S. Army Major Rudy Schroeder's old Lepere mono-prop, biplane which flew to thirty-three thousand feet, breaking the record by over ten thousand feet (or two miles) in February 1920. Schroeder, equipped with oxygen, apparently tolerated the trip well.[3]

Studies by H. G. Armstrong and J. W. Heim in 1937 confirmed Henderson's fears about pressure change with decompression chamber experiments studying high-altitude effects on the middle ear. P. P. Gaursaux and colleagues first learned of the fatal effects of explosive decompressions on laboratory animals in classic chamber experiments of 1938–39.[4] Armstrong detailed the changes in blood gas partial pressures with increasing altitudes between 16,000 and 63,000 feet, the latter known now as "Armstrong's line." At altitudes greater than 5,000 feet, the mental effects of low oxygen could be measured in visual deterioration. At altitudes greater than 10,000 feet, confusion, lethargy, and dizziness were seen. From 35,000 feet to 41,000 feet, there was a severe drop in the arterial oxygen pressure due to an absolute drop in the atmospheric oxygen pressure even if pure oxygen was breathed.[5] At altitudes of 50,300 feet, atmospheric pressure is 86 mm Hg and the partial pressure of oxygen is 18 mm Hg. At such heights, the partial pressure of oxygen in venous blood, which ranges between 45 and 60 mm Hg, is actually greater than in the lungs. Following the concentration gradient, oxygen will simply diffuse out of the blood and into the lungs and be expired, thus ensuring death by oxygen deficiency. Even at 50,000 feet, useful consciousness lasts only six seconds. Since the 1920s, such "heights" were reached only by depressurization in laboratory high-altitude chambers. By the mid-1930s, however, British engineers had developed the Bristol 138A, a single-prop monoplane which flew F. R. D. Swain to 49,945 feet, or more than nine miles up, in a pressurized suit and cabin.[6]

Soon, high-altitude aviation became more of a reality, and during World War II, a matter of survival. To parallel the development of the Hurricane and the Spitfire aircraft by the British, the Germans produced the Messerschmidt class of planes, which were capable of flying to to 35,000 feet in six to ten minutes. The Japanese "Zero" could, by 1941, fly at 19,500 feet at 320 mph. The Mustang (P-51) brought an American presence to higher altitudes, but it was the sea-gull winged Corsair (F4U), capable of a climb rate of 2,265 feet per minute, which finally broke the Japanese hold on the airwar over the Pacific.[7] With wartime German surface-to-air flak ceilings of at least 15,000 feet, high-altitude flying became more of a necessity rather than a theoretical limit when the European air campaign began in earnest, first by the British and then the rest of the Allies in 1942–43. The United States bomber classes B-17 (Flying Fortress) and B-24 (Liberator) practically defined the high-altitude ascents and bombing campaign of the last years of World War II, when flights to 35,000 feet became commonplace.

With a need to become airborne quickly and reach maximal heights, sometimes under enemy fire, the modern air machines of the 1940s subjected the human body to higher altitudes reached at faster rates than had ever been possible. It soon became clear to the United States government, at the outset of the Second World War, that scientists had virtually no physiologic information regarding high-altitude flying. Any successful air campaign would be severely limited in reaching realistic military goals without solid medical and scientific data with which to train pilots and maintain safety.

When it became clear that high-altitude physiology would have enormous implications in the military, Dr. Lewis Weed, chairman of the Medical Sciences Division of the National Research Council, asked the eminent physiologist and historian John F. Fulton, M.D., Ph.D., to establish a committee to look specifically at decompression sickness at high altitude, or "flier's bends."[8] Fulton had gained early notoriety with his research on the physiology of muscle in *Muscular Contraction*, received his doctorate from Oxford in 1925. Two years later, he earned his medical degree from Harvard where he worked under the neurosurgeon Harvey Cushing, whose biography Fulton would later write in 1946. Fulton thereafter concentrated on neurophysiology and was later judged as "one of the world's foremost brain experts."[9] His expertise in electrocardiology, endocrinology, and the physiology of aviation brought an unrivaled academic output, both investigative as well as historical. Fulton's own "bends" experiments were often performed in a decompression-recompression chamber in the basement of the Yale Medical School, where he taught and where students volunteered to be "flown" to thousands of feet in altitude or brought "down" hundreds of feet in pressure. Fulton's Schiff lecture in December 1940 on "Robert Boyle, Father of Aviation Medicine" illustrated his oratory skills, his humor and his passion for the giants of science that came before him. His organizational skills, which helped establish Yale's Cushing Library as a world center for historical medical texts, and his knowledge of physiology made Fulton the optimal coordinator of the national effort to understand the bends. In June 1941, President Franklin Roosevelt had issued an executive order that established the National Research Council (NRC) and the Committee of Military Medicine which was chaired by Fulton's friend Vannevar Bush. Bush, in turn, hired Lewis Weed as chairman of the medical division and he, too was a loyal "Fulton fan." It seemed fitting, then, that Fulton be chosen to chair the NRC's elite council on decompression sickness. It was to be a landmark appointment, for both American aviators of the future as well as workers with compressed air everywhere.

Fulton was first asked by the British Royal Air Force in June 1940 to study the effects of high altitude on monkeys, but was not able to build a chamber for human study in the United States until funding was approved by October 1941. The chamber was, like Haldane's, a huge boiler on its side but the temperature was regulated. Its inner chamber could be heated from -70 degrees Fahrenheit to 70 degrees Fahrenheit in ninety seconds and pressures equivalent to altitudes of forty-five thousand feet produced in three minutes (fig. 41). In meetings with Canadian colleagues Fulton wrote in his diaries of late 1940 of the "growing conviction that it is imperative to pool medical and scientific information for the more successful conduct of the war. The same conviction has grown up in Canada among those responsible such as Sir Frederick Banting and Charlie Best and others."[10]

Fulton compiled a council of the most advanced thinkers in physics and chemistry from the United States' most eminent universities and research laboratories. These scientists were experts in the study of bubble formation and the molecular causes of decompression sickness. The studies had direct application to the medical and physical consequences of high-altitude flight, caisson engineering

Figure 41. Decompression chamber, World War II era. Pictured with a friend, John Farquar Fulton stands in front of Yale's first chamber installed in the late 1930s. Yale's Aeromedical Research Laboratory generated much of the data on high-altitude sickness and bends for aviators in the 1940s. By the 1950s, with jet aviation research taken over by the U.S. Air Force, the program and its chamber were dismantled. (Yale University, Harvey Cushing/John Hay Whitney Medical Library)

and design, and all other environments where a transition from one pressure to a much lower pressure would be done quickly. Many of the leading researchers were at Yale, which had been "fortunate," wrote Fulton in 1945, "to have been associated with the investigative field since its earliest beginnings."[11] Walter Miles, serving on the Committee on the Selection and Training of Aircraft Pilots, studied night vision and developed early infrared goggles later employed by the Army and Navy. Harold Lambert came to Yale to study early "negative gravity" suits to "get ahead" of German designs in the fight against high acceleration forces. Work was also done on the effects of low oxygen pressure on the epinephrine-producing glands, the adrenals, and on spinal cord performance at low pressures. A giant in early research was Leslie F. Nims who was a vital member of Yale's Aeromedical Research Unit (ARU). Among Nims's and the ARU's many accomplishments in decompression research was the development of the finger impedance plethysmograph or finger pulse recording, a staple of every modern operating room which allows moment to moment measurement of arterial pulse wave and its dynamics. Fulton's researchers had at their disposal the newest in laboratory instruments: Plethysmographs, spectrophotometers, chemical analyzers, blood pressure monitors, electrocardiographs, and X-rays. Their work was more sophisticated than any prior worldwide attempts to understand the bends.

The essential event in the bends, as Bert had suspected and Haldane had investigated, was nitrogen bubble formation. If this event could be fully understood, then the bends could be controlled. The committee's assignment, first and foremost, as Weed himself specified, was to "study the problem from a broad biological standpoint." [12] It was also clear that painful symptoms, which rarely arose below twenty-three thousand feet, included almost all of the symptoms seen in divers and tunnel or caisson workers: abdominal cramping, tooth pain, joint discomfort and painful coughing. These symptoms occurred at their most severe usually at altitudes of thirty-five thousand feet. Some unfortunate aviators climbing to high altitude also sustained the life-threatening and dreaded condition of air violently escaping from the lung into the chest cavity: the pneumothorax, or lung collapse.

What was it about bubbles that caused pain? Early observers of bubbles in the blood of decompressed animals suspected that bubbles caused the blockage of blood flow to an affected part, that is, ischemia and the pain it causes. Leslie Nims, senior physiologist earlier at the Brookhaven Laboratory, pointed out, however, that "pain, once developed, can be reduced in intensity or made to disappear from a given region by local pressure applied externally over the site . . . as with a sphygmomanometer [blood pressure cuff]." [13] He also remarked that "some have reported on the relief obtained for pains in the hand and arm by plunging the part into a cylinder of mercury—a very simple means of raising the local tissue hydrostatic pressure. All of these observations," said Nims, "are contrary to the ischemia hypotheses . . . because increasing the anoxia by increasing the occlusion should intensify symptoms rather than diminish them."[14] However, if the pressure of the bubbles is large, a "deformation pressure" is reached and this could result in the bends. Certain tissues, it was thought, had higher deformation pressures and are less likely to manifest symptoms. The ideas of deformation pressure, as espoused by Nims, viewed the bends as a situation in which a bubble deformed nerves in the body to cause pain, and this deformation pressure was proportional to the rate of bubble growth.

What then determined bubble growth? The harmful bubble is a three-dimensional object and in contact with body tissues from where it arises; the bubble is never a perfect sphere. Crucial to bubble growth, however, is the *site* where bubbles first occur, where the first molecule of nitrogen gas comes out of solution to provide a nidus for further accumulation of gas. This process is called "nucleation" and can be seen on the edge of a glass of soda, where long lines of bubbles arise from distinct points of nucleation on the glass walls. Researchers like E. Newton Harvey, professor of physiology at Princeton University, was first able to photograph the actual formation of bubbles in a test tube and various tissues. One tissue that almost never exhibited bubble formation de novo was blood. After prolonged pressure exposures to 100 psi or altitudes of forty five thousand feet he found that "none of the formed elements of blood plays a part in bubble *formation* and that turbulence around the valves of the veins or the heart do not normally *start* bubble formation. . . . Bubbles come, therefore, from nuclei . . . sticking to or formed on or within the endothelial [blood cell] linings of the vascular system and only when they have enlarged to the point of instability do they pass into the blood stream" (italics added).[15] Work by John Bateman, who was chief of the biophysics division of Camp Detrick, an army medical research facility, and served earlier at the Mayo Clinic, established precise formulas and rules which determined bubble growth.[16]

Of course, the presence of bubbles in the blood is not a sine qua non of decompression sickness. In fact, some victims with excruciating pains have no visible bubbles in the blood at all. The mere presence of small bubbles in the blood can be tolerated to a point without any ill effects. Many physician-scientists today think that almost all divers develop *some* microbubbles which develop slowly and pass harmlessly through tissues and then the bloodstream as they are brought slowly to the lungs for expulsion back into the atmosphere.[17] Clearly there was some threshold quantity or location of bubble formation that were required to produce full-blown bends.

Eventually, Fulton's group generated sufficient data on bubble formation that classic Haldanian tables could be further modified to get divers to the surface even faster and safer than ever before. The NRC and ARU work also led directly to the installation of pressurized cabins in aircraft flying routinely above ten thousand feet. At these and higher altitudes, reached easily by most passenger jets of today, decompression due to a broken window or door is so violent that it is termed "explosive." When the pressure drop is severe enough, hypoxia and death can easily occur. Explosive decompressions do form tissue bubbles, but more deadly in these emergency situations is the sudden expansion of the lungs, which causes air embolism and the unstoppable loss of tissue oxygen into the rarified atmosphere.

Orbital flight in the essential vacuum of space is a hazard to any astronaut's body. A sizable piece of debris moving at five miles per second through space can easily smash through the hull of a passing space shuttle, killing any occupants who are not wearing pressurized suits. Most NASA spacecraft are actually pressurized to less than sea level, about 0.8 or 0.9 ATA. Still, an explosive decompression would lower cabin pressure to zero and would result in the "boiling" of the gases dissolved in the astronauts' blood and tissues. This "ebullism" can occur in suborbital flight as well, around sixty-three thousand feet, certainly within the range of most jet fighter, surveillance, and bomber aircraft (fig. 42). There,

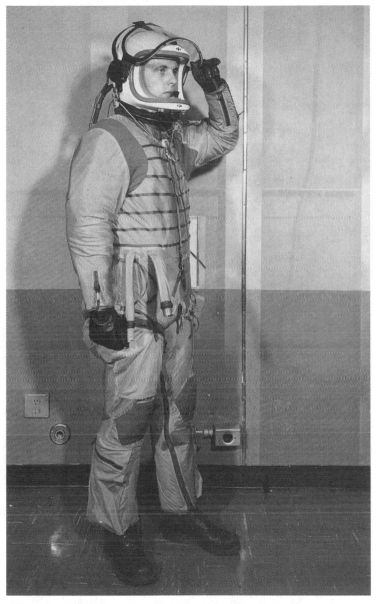

Figure 42. This U.S. Air Force high-altitude pressure suit of 1959 maintained body pressure close to 1 atmosphere when the outside environment was less than one-tenth that. (Courtesy, Raymond Phillips, M.D., Tarrytown, N.Y.)

at a height of twelve miles, death from decompression by accident or bailout is prevented solely by the integrity of the pilot's suit.

By the 1960s and 1970s, ultrasound, nuclear spectroscopy, thermodynamic analysis, and computerized physics programs allowed researchers to put together the final parts of the decompression mystery. The thermodynamic school of thinking, as espoused by Brian Hills, supported the concept that bubble formation, by whatever mechanism, occurs, but that until the pain-provoking threshold is reached, no action need be taken. This led to extremely complex calculations which allowed certain divers to surface immediately if the calculated volume of gas for a given time and depth was not sufficient to cause nerves to become activated and cause pain. This approach predicts a worst-case situation where gas nuclei are activated into a stable phase by decompression but become so intense that the tissue tolerance is exceeded and gas is "dumped" in excess of thermodynamic equilibrium.[18]

Ten major statements, then, can be finally made about the etiology and symptoms of the disease, statements which derived from 100 years of continued research that remains, like all great scientific questions, still active.

1. The primary event of the bends is the activation of nuclei reservoirs of dissolved nitrogen to form bubbles in the body's tissues, mostly fat tissue. The rate of bubble formation depends on the difference between atmospheric and tissue gas pressure and the tolerance of a given tissue for "holding on" to the gas bubbles when decompression occurs.

2. The likelihood of decompression sickness depends upon the number of bubbles, their size, and their position in the body.

3. All sites of important gas bubble formation are extravascular, that is, outside the blood system and in the tissues themselves. Microbubbles that form in the veins are usually harmless as they will be expired during their trip through the lungs. (Explosive de-

compressions, like sudden pressure loss in a passenger jet, may cause dangerous *intravascular* bubbles, i.e., within veins, as the trouble with these is described below.)

4. Gas bubble formation is also influenced by muscle activity. This is similar to a bottle of soda being opened and then jarred by a bump allowing more bubbles to come out of solution. A panicky, muscle-exhausting bailout thus makes matters worse.

5. The huge fat reservoir of gas is "dumped" into the blood of the veins, a harmless occurrence unless the volume is large enough that they coalesce in the right side of the heart. Either this collection of air can cause the right side of the heart to fail or the bubbles can irritate J-receptors, tiny nerve endings in the lungs, which give victims the sensation of painful coughing or "the chokes." This rarely occurs, and bubbles in the veins are thought now to be almost irrelevant to routine bends accidents. Minor symptoms from venous bubbles cause such minor complaints (like itching) that they are referred to as "the niggles," in marked contrast to the events below.

6. If bubbles escape exhalation in the lungs and are massive enough to bypass through pulmonary vessels they will arrive in the left side of the heart. From here, they can be ejected to any place in the body but most seriously to the brain. Accumulating in the tiny watershed vessels of the fragile nervous tissue of the brain, they slowly block the supply of oxygen, causing seizures, stroke, and irreversible brain damage or death. About 30 percent of the population have tiny holes that abnormally connect the blood collecting cavity (the atrium) of the right side of the heart with the atrium on the left side of the heart. The defects allow bubbles to escape from the right side *before* going to the lungs and arrive immediately in the left side of the heart, where they can then be sent out to the body and the brain.[19] Such people have a very high risk of serious bends in compressed air diving.[20]

7. Spinal cord bends usually occur after absorbing nitrogen

amounts that are much greater than those which can cause "limb bends." The spinal cord is amazingly fragile. Nerves have very limited tolerance for decreased blood supply (arterial problems) or decreased blood drainage out of the cord (capillary and venous problems).

8. The delay in developing pain after decompression is due to the time required to "grow" enough bubbles to stimulate nerve endings in the various tissues.

9. Nerve-pain thresholds can be permanently altered by some mechanism of blood-flow related damage so that "bends" pain can occur long after all the gases have been removed from the body. This explains why some patients continue to have pains in their limbs or muscles even years after an accident. Correspondingly, some victims can develop a tolerance to the same gas load that formerly had given them pain.

10. Bone disease occurs because of "watershed" blood flow areas of susceptible bones (the femoral and humeral heads) which, like the spinal cord, tolerate venous bubbles poorly. The bone-forming cells, deprived of oxygen and facing increased pressure in a non-expandable bone matrix, can die, leaving abnormal bone formation behind. The strain of working in tunnels, caissons, and other underwater conditions, coupled with the tissue strain of multiple decompressions and tissue-insults can cause cartilaginous and bony injury. Sometimes these injuries result in full-blown cases of permanent bone damage called dysbaric osteonecrosis months to years after the actual decompression accident.

The modern concepts of decompression sickness are becoming more and more refined with the increased ability to focus on the primary event, bubble formation, now with magnetic resonance imaging, three-dimensional X-ray and computed tomography, and microcatheter tissue gas sampling. Ultrasound has been now able to identify in the blood vessels of the eye the minute bubbles that all divers suspect occur from decompression at any depth but which never cause symptoms. Highly sophisticated techniques will be able eventually to fill the remaining blanks in to the point that the laws of gas behavior can be predicted anywhere, from the bottom of the Mariana Trench at thirty-five thousand feet of sea water to the vacuum of space.

# Notes

## Chapter 1. Introduction

1. Engel, *The Sea*, 10–11.
2. Doll, "Decompression Sickness," 3, 14. Clenney and Lassen, "Recreational Scuba Diving Injuries," 1761.
3. Dugan, *World Beneath the Sea*, 181–182.
4. Kylstra, "Liquid Breathing and Artificial Gills," 159.
5. Thackrah, *Arts, Trades, and Professions*, 82.
6. Miller, "A Production of Amino Acids Under Possible Primitive Earth Conditions," 528–529.
7. Urey, "On the Early Chemical History of the Earth and the Origin of Life," 351–363.
8. Gilbert, "Evolutionary Aspects of Atmospheric Oxygen and Organisms," 1063–1064.
9. Ibid., 1068–1070.
10. Gould, *Wonderful Life*, 94.
11. Zapol, "Diving Physiology of the Weddell Seal," 1049.

## Chapter 2. The Discovery of the Atmosphere

1. Sigerist, *History of Medicine*, vol. 1, 348.
2. Ibid., 33, 263.
3. Parker, *Miasma*, 33, 308–321. The English form of the word *miasma* loses the Greek meaning of metaphysical infection of evilness or perfidy and reflects a more modern concept of infection. "Bad air" became known to Europeans as malaria and today specifies a parasitic infection with the genera *plasmodia*. Malaria still remains the world's number one cause of death owing to infection.
4. Sigerist, *History of Medicine*, 89–91. Bowra, *Classical Greece*, 184.
5. Sigerist, *History of Medicine*, 91.
6. Ibid., 262–263, 278.
7. Ibid., 322–323.
8. Boorstin, *The Discoverers*, 404–405. Before he turned 26 years old, Newton

had developed his theories on the binomial theorem, the laws of refraction and light, the laws of motion and of gravity and the groundwork for calculus. As he himself had oft been quoted to say, "I do not know what I may appear to the world, but to myself I seem to have been only like a boy playing on the seashore and diverting myself in now and then finding a smoother pebble or a prettier shell than ordinary, whilst the great ocean of truth lay undiscovered before me."

9. Ibid., 403.

10. Lapp, *Matter*, 63.

11. Proctor, "History of Breathing Physiology," 21.

12. Bottazzi, *Leonardo da Vinci*, 377. Baroni refers "the element of fire continually consumes the air, which in part nourishes it, and would remain in contact with the vacuum if the surrounding air did not rush in to fill it" to Cod. Atl., fol. 237 v.

13. Proctor, "History of Breathing Physiology," 41–42.

14. Ibid., 67–68.

15. Ibid., 70.

16. Ibid., 173. Mayow, *Medico-Physical Works*, 201. John Mayow (1643–1679), a graduate of Oxford, was befriended by Robert Hooke, and was elected a Fellow of the Royal Society in 1678. Mayow came very close to identifying oxygen as the vital component needed for life but called it a "nitro-aerial spirit" thus refuting the then-accepted theory that breathing served to cool the heart. Mayow was the first to clearly describe the relationship between the process of breathing, the lungs, and the atmosphere, something even Leonardo failed to grasp. "Hence it is that when the inner sides of the thorax . . . are drawn outwards by muscles whose function it is to dilate the chest, and the space in the thorax is enlarged, the air which is nearest the bronchial inlets, now that every obstacle is removed, rushed under the full pressure of the atmosphere into the cavities of the lungs, and by inflating them occupies the space of the expanded chest."

17. Magnus, *Historia om de Nordiska Folken*, 9–11.

18. Ramazzini, *De Morbus Artificium Diatriba*, i–ii.

## Chapter 3. The Sea of Air Around Us

1. Sigerist, *History of Medicine*, 109–111.

2. Proctor, "History of Breathing Physiology," 80.

3. Ibid.

4. Ibid.

5. Ibid., 81. A torr is a unit of pressure, named after Evangelista Torriccelli

(1608–1647), which is enough to support a millimeter of mercury in a barometer. One atmosphere is 760 torr or 760 mm Hg. Meterologists will, for this reason, refer to changes in pressure as changes in torr or mm Hg. In 1971, torr was formally replaced in academic settings by the pascal ($N/m^2$), which is equal to 0.0075 torr.

6. Hudson and Nelson, *University Physics*, 366.
7. *Encyclopaedia Britannica*, 1992, s.v. "vacuum technology," 228.
8. Faires, *Thermodynamics*, 7, 9.
9. Ibid., 9.
10. Cobb and Goldwhite, *Creations of Fire*, 202–203, 207.
11. Bernal, *Social Sciences in Science and History*, 462.
12. Proctor, "History of Breathing Physiology," 102. Also found in Thomas Birch's *Works of the Honourable Robert Boyle*.
13. Ibid., 129.
14. Ibid., 130–131.
15. Ibid., 209.
16. Ibid.
17. Cobb and Goldwhite, *Creations of Fire*, 158.
18. Empleton, *New Science of Skin and Scuba Diving*, 39–40.
19. Lavoisier's hypotheses were correct, but, given the analytical equipment of history, he could not gravitate the exact proportions of air's different constituents. His 1:3 ratio comes close to describing the 1:4 oxygen-nitrogen ratio known today.
20. *Dictionary of Scientific Biography*, s.v. "William Henry."

## Chapter 4. The Pneumatic Revolution

1. Resnick and Halliday, *Physics*, 368, appendix 31.
2. Hudson and Nelson, *University Physics*, 366. For any spherical container with a radius R partially evacuated of air, the weight (force, F) on its surface can be readily calculated with the formula F=p × (R) 2 × (pressure difference). A conservative estimate is that von Guericke was able to create a vacuum of 0.1 ATA (1.47 psig) between two hemispheres creating a sphere two feet wide. The formula is thus F=p × (12) 2 × 13=5,980 pounds. Von Guericke also knew that *one* team of eight horses pulling the sphere attached to a tree would have produced the same effect as two teams of eight, but far less dramatically.
3. Resnick and Halliday, *Physics*, 364–369. The greatest vacuum was achieved by K. Odaka and S. Ueda of Japan in January, 1991 in a stainless steel chamber. Using ultra-efficient valve mechanisms, hydraulics and pump

power, the two barophysicists recorded a pressure of $7.0 \times 10^{-16}$ atm or 70 quadrillionths of an atmosphere (70 femtobars). Using Avogadro's number ($6.023 \times 10^{23}$ molecules/mole and 1 mole $= 22$ L gas) the "near perfect" Odaka-Ueda vacuum still contained 30,000 gas molecules per milliliter. *Guinness Book of World Records*, 1996, s.v. "Vacuum, world's greatest."

4. Wilson, *American Science and Invention*, 50.

5. Bernal, *Social Sciences in Science and History*, 577.

6. Wilson, *American Science and Invention*, 51.

7. Bernal, *Social Sciences in Science and History*, 581.

8. Wilson, *American Science and Invention*, 51–52. Units of work are difficult to extrapolate from one engine to another. However, Newcomen's Atmospheric Engine established what was then known as an "engine's duty": a half-million foot-pounds of work per bushel of coal (about 20 kilograms) burned or about 10 horsepower. Watt's changes quadrupled the "engine duty," and nineteenth-century engines would not do much better until the expansile energy of steam was replaced by the energy source that defined the twentieth century: gasoline. Samuel P. Langley's radial gasoline engine of 1903, for example, developed 52 horsepower. By 1942 Daimler-Benz built a coupled engine capable of 3,100 horsepower for use on the Heinkel long-range bomber of the German Luftwaffe.

9. Hoff, *Bibliographic Sourcebook*, 7–10.

10. Dugan, *World Beneath the Sea*, 30.

11. Hamilton-Paterson, *Great Deep*, 166–167.

12. Bachrach, "Short History of Man in the Sea," 2.

13. Ringqvist, "Ventilatory Capacity in Healthy Subjects," 1–3.

14. Stigler, "Kraft unserer Inspirations Muskulatur," 235. A friend tells the story that, as a boy, he dove off a raft with a standard three-foot-long garden hose and attempted to breathe through it while his pals held the other end above the surface. Inhaling with all his might, he could not get any air. After several seconds of attempts, he surfaced in a panic—convinced that his friends had actually plugged the tube with their fingers—only to find them innocently holding the tube open just as he had left it.

15. Lundgren, "Gas Physiology in Diving," 1002.

16. Dugan, *World Beneath the Sea*, 40, 43–45.

17. Lundgren, "Gas Physiology in Diving," 1003–1004.

18. Davis, "Deep Diving and Underwater Rescue," 1032–1034, 2–3.

19. Kindwall, "Short History of Decompression Sickness," 1–2.

20. Hoff, *Bibliographic Sourcebook*, 9.

21. Bachrach, "Short History of Man in Sea," 7.

22. Dugan, *World Beneath the Sea*, 34, 38.
23. *Encyclopaedia Britannica*, 1992, s.v. "John Smeaton," 267.
24. Kindwall, "Short History of Decompression Sickness," 2–3.

Chapter 5. Triger's Caisson

1. *U.S. Navy Diving Manual* (1959), 17. 33 feet of sea water or 34 feet of fresh water exerts a pressure equal to 14.7 psig.
2. Wilson, *American Science and Invention*, 51.
3. Jeff Ehmen of NASA, E-mail to author, June 20, 1996. These huge F-1 engines needed to generate 7.5 million pounds of thrust to propel the Saturn V rocket into space. The brakes for the turbopumps which drew the fuel and oxidizer from their gas tanks into the stage could generate up to 55,000 horsepower *each*.
4. Faires, *Thermodynamics*, 99–100, 421–430.
5. Bernal, *Social Sciences in Science and History*, 592–593. Conversation with engineer of Essex steam train #103, 7 July, 1996. The power of an engine is never the absolute determinant of work but rather the application of power. A graphic example is that of most nineteenth century locomotives, which could explode if pressures in the water boilers went much over 30 pounds per square inch gauge. However, by applying as little as 10 psig to these trains' two or more enormous cylinders, some almost 30 inches in diameter, the effective surface area is very large. Suddenly, 10 psig, or less than what a diver needs to breathe comfortably at 23 feet of depth, can create a force of 50,000 pounds and can move a 400-ton train. A pressure of 6 psig can keep it going at 30 mph.
6. Caillaux, "Notice sur la Vie et les Travaux de M. Triger," 547–559. Barral, "Triger," 271–273. Triger's first name has been long sought after by diving historians. Triger is partly to blame. In none of his correspondence with the engineering societies of mid-nineteenth-century France, does he give his first name. One report by Georges Barral of the 1880s, a who's who of nineteenth-century French engineering, lists Triger as Charles-Jean born January 11, 1800, and dying suddenly on June 30, 1872. Alfred Caillaux wrote Triger's obituary in 1868 in the Bulletin of the Geologic Society of France, reported him as Jules Triger born March 11, 1801, and dying, again, suddenly, on December 16, 1867, while attending a meeting of the Société géologique de France. Since both biographies describe C.-J. and J. Triger as the first to use compressed air in a caisson design, there is little doubt that they were describing the same man. Barral wrote that C.-J. Triger

"established a superior apparatus which established a pocket of air and a compressed air atmosphere by means of insufflation." Caillaux praised Jules Triger, writing, "It was truly an epochal event when he led the creation of a new apparatus, well engineered and of great utility in the art of mining, which became the object of lofty rewards: the application of compressed air for the drilling of a well." Regardless of Triger's christian name, his mark on engineering and diving is truly unmistakable.

7. Glossop, "Jules Triger, 1801–1867," 538–539, and also Caillaux, "Notice sur la Vie et les Travaux de M. Triger," 548. Barral, "Triger," 271.

8. Glossop, "Triger," 548.

9. Triger, Letters, April 14, 1834.

10. West, *Innovation and the Rise of the Tunnelling Industry*, 150–151.

11. Triger, "Emploi de l'Air Comprimé," 233–234, and Triger, *Minutes*, Assemblée Générale, 1841–1847.

12. Triger, "Mémoire sur un Appareil," 91–92.

13. Pol and Watelle, "Mémoire sur les Effets," 258. Triger's men noticed less breathlessness or "moins essoufflés" while going up the ladder in the caisson compared to such exhausting work on the surface.

14. Smith, *Effects of High Atmospheric Pressure*, 5.

15. Pol and Watelle, "Mémoire sur les Effets," 250–259, 265–270, 278.

16. Prade, M.-F., E-mail to J.L.P., June 16, 1996.

17. Pol and Watelle, "Mémoire," 241.

18. Flexner, *Medical Education*, 2–10. On seeing the standardization of medical training throughout the world fueled by laboratory-based knowledge and research, he claimed that "science, once embraced, will conquer the world."

19. Bordley and Harvey, *Two Centuries of American Medicine*, 41–45.

20. Ibid., 132, 157–159.

21. François, "Des Effets sur les Ouvriers Travaillant," 319.

22. Pol and Watelle, "Mémoire," 276 ("Non numerandae sed perpenendae observationes").

23. Ibid., 249–250: "outre l'impossibilité de siffler, notée par M. Triger à partir de 3 atmosphères."

24. Smith, *Effects of High Atmospheric Pressure*, 6.

25. Ibid., 277–278.

26. Ibid., 261.

27. Ibid., 277–278.

28. Hoff, *Bibliographic Sourcebook*, 260.

29. François, "Des Effets sur les Ouvriers Travaillant," 300–302.

30. Ibid., 318.

31. Ibid., 289–290.
32. Smith, *Effects of High Atmospheric Pressure*, 10.
33. François, "Des Effets sur les Ouvriers Travaillant," 319.
34. Ibid.

Chapter 6. James B. Eads and the St. Louis Bridge

1. Primm, *Lion of the Valley*, 299–300.
2. Ibid., 304.
3. Ibid., 305–306. Woodward, *History of the St. Louis Bridge*, 240–245.
4. Primm, *Lion of the Valley*, 305.
5. Woodward, *History*, 246.
6. Ibid., 247.
7. Smith, *Effects of High Atmospheric Pressure*, 11.
8. Woodward, *History of the St. Louis Bridge*, 247.
9. Primm, *Lion of the Valley*, 306.
10. Temple, *Genius of China*, 58.
11. Ibid.
12. *Encyclopaedia Britannica*, 1992, s.v. "Public works," 339.
13. Ibid., 347.
14. Ibid., 348.
15. Temple, *China*, 58.
16. McCullough, *Great Bridge*, 63.
17. *Encyclopaedia Britannica*, 1992, s.v. "Public works—bridges," 335. All suspension bridges assume the shape of a catenary curve. It is a simple, yet geometrically beautiful shape assumed by a free-lying rope or cord hung between to points in space described by the equation y = k × cosh × (x/k). Deriving from the Latin *caten*, chain, it is very stable if the "chain" is stiff as seen in the wire rope cables of Roebling's bridge. If solid in form, the curve can be built upside down and is equally strong, illustrated by the famous example of the Gateway Arch in St. Louis built in 1975, the world's tallest free-standing, upside-down catenary curve.
18. Ibid.
19. Woodward, *History of the St. Louis Bridge*, 246–247.
20. Based on a 60% efficient output of an ideal, isoentropic, reciprocating air compressor supplying 50,000 cubic feet at 40 psig.
21. Woodward, *History of the St. Louis Bridge*, 248.
22. Ibid., 254.
23. McCullough, *Great Bridge*, 186.

24. Woodward, *History of the St. Louis Bridge*, 247.
25. Behnke, "Decompression Sickness Following Exposure to High Pressures," 61.
26. Ibid., 63.
27. *Association of Diving Contractors Therapy Guidelines*, 4–5.
28. Woodward, *History of the St. Louis Bridge*, 248.
29. Ibid.
30. Ibid., 248–249.
31. Ibid., 249.
32. Ibid., 251.
33. Kindwall, "Optimum Schedules for Caisson Decompression," 157.
34. Thackrah, *Arts, Trades, and Professions*, 90.
35. Woodward, *History of the St. Louis Bridge*, 250.
36. Ibid.
37. Hill, "Decompression Sickness," appendix i.
38. Woodward, *History of the St. Louis Bridge*, 254.
39. Kindwall, "Optimum Schedules for Caisson Decompression," 13. Walder, "The Prevention of Decompression Sickness," 459–460.
40. Jaminet, *Physical Effects of Compressed Air*, 116–177.
41. Woodward, *History of the St. Louis Bridge*, 252.
42. Jaminet, *Physical Effects of Compressed Air*, 116.
43. *U.S. Navy Diving Manual*, "Air Decompression, Table 1–5" (1970).
44. Woodward, *History of the St. Louis Bridge*, 253.
45. McCullough, *Great Bridge*, 186, where he reports sixteen having died from DCS during Eads's bridge construction.
46. Woodward, *History of the St. Louis Bridge*, 253.
47. Ibid., 258.
48. Ibid., 257, and see Moir, *Tunneling*, 574.
49. Ibid., 258.
50. Woodward, *History of the St. Louis Bridge*, 259–260.
51. Jaminet, *Physical Effects of Compressed Air*, 113–114.
52. Ibid., 116.
53. Woodward, *History of the St. Louis Bridge*, 258.
54. Wilson, *American Science and Invention*, 88–89.

Chapter 7. The Roeblings and the Brooklyn Bridge

1. From data courteously supplied by the South Street Seaport Museum of New York and the Merseyside Maritime Museum of Liverpool, England. Liverpool served as the single most important sailing hub linking Europe with the New

World of the United States, Latin America and beyond. With the advent of the train, New York rose to worldwide prominence. During the year 1865, 455 steam and 4,271 sailing vessels sailed into the port. Data from 1866 showed that steam vessels into Liverpool numbered 1,328, while sailing vessels numbered 3,383. The following year, Liverpool's foreign traffic had a net decrease of 15%, whereas by 1877 New York docks claimed a 24% increase in traffic, to 6,224 incoming vessels.

2. *Cambridge Biographical Dictionary*, s.v. "Roebling, John A."
3. McCullough, *Great Bridge*, 61.
4. McCullough, *Great Bridge*, 42.
5. Ibid., 70–71.
6. Ibid., 563–564.
7. Opisthotonos, Greek: *opisthon*, drawn behind, + *tonos*, spasm or seizure.
8. Ward, *Civil War*, 308–309.
9. Catton, *Civil War*, 466–467.
10. McCullough, *Great Bridge*, 145.
11. Ibid., 212.
12. Ibid., 174, 564.
13. Ibid., 193, 292.
14. Ibid., 201.
15. Ibid., 295.
16. Smith, *Effects of High Atmospheric Pressure*, 15.
17. Andrew Heermance Smith was the first physician to propose the use of a medical lock for the treatment of the bends. According to the New York Academy of Medicine, he was also the first to propose the use of gaseous oxygen for general clinical lung disease, twenty years before the proposals by England's J. S. Haldane. New York Academy of Medicine Archival Biography.
18. Smith, *Effects of High Atmospheric Pressure*, 18–19.
19. Ibid., 76–77.
20. Hong, "Mixed-Gas Saturation Diving," 1038–1040.
21. Keenan, "Hormones of the Hypothalamus and Pituitary Gland," 866.
22. Smith, *Effects of High Atmospheric Pressure*, 23–24.
23. Ibid., 24.
24. McCullough, *Great Bridge*, 318–319.
25. Smith, *Effects of High Atmospheric Pressure*, 15, 39–43.
26. Ibid., 30.
27. Ibid., 31.
28. Ibid., 32.
29. Ibid.

30. Ibid., 41–42.
31. McCullough, *Great Bridge*, 318.
32. Ibid., 49–50.
33. McCullough, *Great Bridge*, 315–317.
34. Ibid., 564.
35. Bordley and Harvey, *Two Centuries of American Medicine*, 163.
36. Wilson, *American Science and Invention*, 216–217.

## Chapter 8. Tunneling Underground and Underwater

1. Moir, "Tunnelling," 568.
2. Black, *Story of Tunnels*, 3–4.
3. Kindwall, "Optimum Schedules for Caisson Decompression," 3.
4. Diehl, *Late, Great Pennsylvania Station*, 34.
5. Black, *Story of Tunnels*, 95–96.
6. Ibid., 23–29.
7. Ibid., 23.
8. Ibid.
9. West, *Innovation and the Rise of the Tunnelling Industry*, 150–151.
10. Black, *Story of Tunnels*, 31–32.
11. Ibid., 62–63.
12. Moir, "Tunnelling," 574.
13. Black, *Story of Tunnels*, 63.
14. Hewett and Johannesson, *Compressed Air Tunnelling*, 300.
15. Ibid., 131–132.
16. Moir, "Tunnelling," 573.
17. Ibid., 574.
18. Hewett and Johannesson, *Compressed Air Tunnelling*, 429.
19. Moir, "Tunnelling," 574.
20. Ibid.
21. Bordley and Harvey, *Two Centuries of American Medicine*, 43–44. Moir's recompression was not homeopathy in the traditional sense. The founder of homeopathy, Samuel C. Hahnemann (1755–1843), found that when he took large doses of quinine he developed a pounding heart, fevers, and sweats. He recalled that these symptoms were similar to malaria, which can often be cured by quinine (now known to be a weak antibiotic). Hahnemann thought quinine cured the disease, however, by recreating its symptoms. He reasoned that medicines should harmlessly reproduce the symptoms of the disease one is attempting to cure which he dubbed as "similia similibus curantur" (likes are cured by likes). Moir's use of recompression was not

homeopathic in the traditional sense, therefore, since compressed air exposure has no symptoms and recompression ameliorates, not re-creates, the pains of the bends.

22. Moir, "Tunnelling," 574.
23. Black, *Story of Tunnels*, 191.
24. Haxton and White, "Compressed Air Environment," 12.
25. Moir, "Tunnelling," 574.
26. Diehl, *Late, Great Pennsylvania Station*, 48–49.
27. Black, *Story of Tunnels*, 108.
28. Ibid., 103–104.
29. Diehl, *Late, Great Pennsylvania Station*, 63.
30. Hewett and Johannesson, *Compressed Air Tunnelling*, 431. Levy, "Workers in Compressed Air," 73.
31. "An Appreciated Inventor," *Literary Digest* (1908) 36: 593, 1908. Moir returned to England, where his career in tunneling furthered his growing reputation for skill, ingenuity, and humanitarianism. During the First World War, he oversaw the purchasing and manufacturing of most of the steel used for railroad commerce and construction throughout Europe. For these efforts and for his life-long commitment to engineering excellence and worker health he was knighted by King George V in 1916. He remains one of the few engineers who was a full member of both the Royal Society of Engineers in Britain and of the American Society of Civil Engineers in the United States. "Sir Ernest William Moir" (obituary), *The Engineer*, June 23, 1933, 629.

Chapter 9. Paul Bert and the Cause of Decompression Sickness

1. Haldane and Priestley, *Respiration*, 252. The ichthyologist Richard Pyle routinely dives to 100–150 meters off Hawaii to collect fish. Pyle, not unlike his Mediterranean predecessors, uses a hypodermic needle to puncture the fish bladder prior to collection and ascent to the surface. Mike Busuttili, http://www.divernet.com/usrep696.htm.
2. Bernal, *Social Sciences in Science and History*, 555.
3. Ibid., 589–590.
4. Ibid., 568–569.
5. Fulton, *Decompression Sickness*, i–ii.
6. Bert, *Barometric Pressure*, 860, 861, 863.
7. Ibid., 868–869.
8. Dalton, *Human Physiology*, 245–246.
9. Bert, *Barometric Pressure*, 512.
10. Ibid., 459, 501, 881.

11. Ibid., 881.
12. Ibid., 884.
13. Ibid., 884–885.
14. Ibid., 887.
15. Smith, *Effects of High Atmospheric Pressure*, 33.
16. Ibid., 34–35.
17. Empleton, *New Science of Skin and Scuba Diving*, 39–40.
18. *U.S. Navy Diving Manual* (1959), 25. Henry's Law if that the amount of a gas that will dissolve in a liquid at a given temperature is almost directly proportional to the partial pressure of the gas. Henry wrote "almost" because he did not include gas solubility in his studies.
19. Boycott and Damant, "Experiments on the Influence of Fatness," 445.
20. Smith, *Effects of Atmospheric Pressure*, 37.
21. Work on the Clyde tunnel in 1959 showed that the skin fold thickness over the triceps and below the shoulder blade correlated closely with the likelihood of developing the bends. Curiously, waist girth does not correlate as well (Eric P. Kindwall, M.D., personal communication).

Chapter 10. John Scott Haldane and Staged Decompression

1. *Dictionary of Scientific Bibliography*, s.v. "John Scott Haldane."
2. Guyton, *Medical Physiology*, 511–512.
3. Burton, "Hazardous Materials," 766.
4. Chinard, "Priestley and Lavoisier," 213–215, 220.
5. Boycott, Dament, and Haldane, "Prevention of Compressed Air Illness," 348.
6. Haldane and Priestley, *Respiration*, 334.
7. Ibid., 334.
8. Boycott and Damant, "Experiments on the Influence of Fatness," 447.
9. Ibid., 455.
10. *U.S. Navy Diving Manual* (1959), 193.
11. Boycott et al., "Prevention of Compressed Air Illness," 354.
12. Ibid., 355.
13. Haldane and Priestley, *Respiration*, 341. Haldane's essential belief was that bubbles did not form until the state of "supersaturation" was overcome. This thesis, however erroneous, supported a truly laboratory-based approach to DCS prevention that defined research in the bends for the next six decades: "We can thus regulate the rate of decompression so that no part of the body is at any time supersaturated to such an extent as to cause risk of bubble formation." Also, see Singstad, "Industrial Operations in Compressed Air," 522.

14. Boycott et al., "Prevention of Compressed Air Illness," 357.
15. Ibid., 424–425.
16. *U.S. Navy Diving Manual* (1970), tables 1–5.
17. Haldane and Priestley, *Respiration*, 352. "It is probable that the bubbles first formed in supersaturated blood and tissues are extremely small and comparatively harmless. One can observe the formation of these minute bubbles in water which has stood in a pipe under pressure in contact with air. When the tap is opened the water comes out milky with minute bubbles, but no large bubbles present. The smallness of the bubbles leaves time to deal with accidental cases of sudden decompression."
18. Dugan, *World Beneath the Sea*, 38–40.
19. Macinnis, "Open-Sea Diving Techniques," 24–28.
20. *U.S. Navy Diving Manual* (1959), 3.

Chapter 11. Decompression Sickness and the Government

1. Burchell, *Age of Progress*, 127–137.
2. Thackrah, *Arts, Trades, and Professions*, 111.
3. Author's count of newspaper covers listed in text.
4. *Scientific American* 22 (1870), 357.
5. Rosen, *History of Public Health*, 420.
6. Carter, "Role of Government in Preventing Ill Health at Work," 15.
7. Ramazzini, *De Morbus Artificium Diatriba*, iii–iv.
8. Ibid., v
9. Thackrah, *Arts, Trades, and Professions*, 90.
10. Ibid., 64. He also wrote, "We remark with regret the men's inattention to health, their indifference to the prevention of disease. They think nothing of injurious agents till their health is destroyed and the time for prevention is passed" (Ibid.).
11. Rosen, *History of Public Health*, 421–424.
12. Ibid.
13. Ibid.
14. West, *Innovation and the Rise of the Tunnelling Industry*, 138–140.
15. Singstad, "Industrial Operations in Compressed Air," 518.
16. Hewett and Johannesson, *Compressed Air Tunnelling*, 422.
17. Ibid.
18. Ibid., 431.
19. Hoff, *Bibliographic Sourcebook*, 263.
20. Ibid., 263–264.
21. Langlois, "La prophylaxie des accidents dans l'air comprimé," 55.

22. Hewett and Johannesson, *Compressed Air Tunnelling*, 430.
23. The great tunnel engineers Hewett and Johannesson lamented the poor compliance of many workers back in the days of the Hudson river tunnels writing "the great trouble with . . . rules made for the benefit of the men is that they seem to look upon them as assaults upon their personal liberty and spend much ingenuity in evading or breaking the rules" (Ibid., 425).
24. Levy, "Workers in Compressed Air," 73.
25. State of New York, Industrial Code Bulletins, "Work in Compressed Air," 7.
26. Hoff, *Bibliographic Sourcebook*, 269. Still, Keays and Japp's recommendations led to the New York State 1909 laws which required contractors by law to provide physicians, medical locks, screening physical exams and working conditions. The laws also classified a violation of these statutes as a misdemeanor punishable with fines and/or imprisonment. *State of New York Labor Bulletin* (1910), 220.
27. Hewett and Johannesson, *Compressed Air Tunnelling*, 433.
28. Levy, "Workers in Compressed Air," 73.
29. State of New York, *Annual Report of the Industrial Commissioner, 1922*, 144.
30. State of New York, Industrial Code Bulletins, 8.
31. Hewett and Johannesson, *Compressed Air Tunnelling*, 431.
32. Ibid.
33. Singstad, "Industrial Operations," 517.
34. Hewett and Johannesson, *Compressed Air Tunnelling*, 436.
35. Ibid., 436–437.
36. Black, *Story of Tunnels*, 188–189.
37. Obituary, *New York Times*, October 14, 1937, column 2.
38. Singstad, "Industrial Operations," 522.
39. Hoff, *Bibliographic Sourcebook*, 156–157.
40. Bassoe, "The Late Manifestations of Compressed Air Disease" (1913): 527, 532, 535.
41. Ibid., 536.
42. Ibid., 539.
43. Kahlstrom et al., "Aseptic Necrosis of Bone," 129–130, 144.
44. West, *Innovation and the Rise of the Tunnelling Industry*, 148.
45. Kindwall, "Optimum Schedules for Caisson Decompression," 160.
46. Kindwall, "Medical Aspects of Compressed Air Tunnelling," 169.
47. Kindwall, "Optimum Schedules for Caisson Decompression," 160.
48. Kindwall, "Medical Aspects," 171. At an Undersea Medical Society conference on July 25, 1979, the great barophysiologist Albert Behnke summed up his frustration with the OSHA table inadequacies saying, "In our experi-

ence in the BART project [Bay Area Rapid Transit], at a pressure level of 30 psig, . . . there was an incidence of decompression sickness which was in the range from 13–55%. At a given time we had as many as eight men under treatment . . . I think we have enough evidence to indicate again that in the pressure range of 30 to 35 psig the tables are totally inadequate." Undersea Medical Society, workshop report to NIOSHA July 25, 1979, 3.

49. Kindwall, "Optimum Schedules for Caisson Decompression," 160.
50. Ibid., 162.
51. Ibid., 157.
52. Ibid., 165.
53. Bert, *Barometric Pressure*, 884.
54. Kindwall, "Optimum Schedules for Caisson Decompression," 166.
55. Ibid., 167.
56. Yodaiken, memorandum, August 2, 1988, 2, 3.

## Chapter 12. Diving Records and Rescues

1. Singstad, "Industrial Operations in Compressed Air," 499.
2. Bachrach, "Short History of Man in the Sea," 7.
3. Dugan, *World Beneath the Sea*, 35.
4. Ibid., 45.
5. Ibid., 45–46. Using Fleuss's rebreathing apparatus, Alexander Lambert became one of the most famous divers of the 1800s. In 1880, he went into the absolute darkness of a flooded tunnel under the Severn River in England to close a blocked air-lock door. In 1883, he repeated the feat, but Fleuss's apparatus failed him and Lambert almost died from carbon dioxide poisoning. Finally, while salvaging millions of dollars in silver from the sunken *Alphonso XII*, he developed the bends and this forced his permanent retirement *U.S. Navy Diving Manual* (1959), 2.
6. Ibid., 3.
7. Ibid.
8. Kindwall, letter to author, July 30, 1996.
9. Thomson, "Helium in Deep Diving," 36–38. Bühlmann, "Decompression Theory: Swiss Practice," 349–350.
10. McLennan, "Helium: Its Production and Uses," 778–779.
11. Hoff, *Bibliographic Sourcebook*, 290–291.
12. Kindwall, "Short History of Decompression Sickness," 5.
13. *U.S. Navy Diving Manual* (1959), 5–7.
14. Kindwall, "Short History of Decompression Sickness," 5.

15. Kindwall, "Treatment of Decompression Sickness," 5.
16. Dugan, *World Beneath the Sea*, 48. Bachrach, "Short History of Man in the Sea," 5.
17. Kindwall, "Short History of Decompression Sickness," 7.
18. Ibid.
19. Singstad, "The Queens Midtown Tunnel. Discussion," 375–386.
20. Kindwall, "Short History of Decompression Sickness," 7.
21. Ibid., 7–8.
22. Ibid.
23. Lee, "Therapeutic Effects of Different Tables," 11.
24. Kindwall, "Decompression Sickness," 11.
25. Beyerstein, "Why Another Therapy Scheme?" 4.
26. Ibid.
27. George, "Functional Adaptations of Deep-Sea Organisms."
28. Kindwall, "Short History of Decompression Sickness," 7–8.
29. Bennett, "The High Pressure Nervous Syndrome: Man," 238–239.
30. Hong, "Mixed-Gas Saturation Diving," 1027.
31. Ibid., 1026.
32. Kindwall, "Short History of Decompression Sickness," 8.
33. P. B. Bennett, personal communication.
34. Bennett, "Inert Gas Narcosis," 214.
35. Hong, "Mixed-Gas Saturation Diving," 1026–1027. The current record for a saturation dive using hydrogen is held by Arnaud de Nechaud de Feral of Comex who from October 9 to December 21, 1989, stayed at the equivalent of a depth of 985 feet(*Guinness Book of World Records*, 1996, s.v. "Diving record, saturation").
36. Hong, "Mixed-Gas Saturation Diving," 1031.
37. Ibid., 1032. The divers were Stephen Porter, Len Whitlock, and Erik Kramer.
38. Ibid.
39. Dugan, *World Beneath the Sea*, 181–182.
40. DuBois et al., "Alveolar Gas Exchange During Submarine Escape," 509.
41. E. P. Kindwall, M.D., personal communication. The longest saturation dive in an undersea dwelling was by Richard Presley of Key Largo, Florida. Entering the Project Atlantis underwater habitat on May 6, 1992, he surfaced on July 14, after almost 70 days of breathing compressed air at a depth of 60 feet. He took about three days to completely decompress. *Guinness Book of World Records*, 1996, s.v. "Diving, world's longest saturation."
42. MacDonald, "Hydrostatic Pressure Physiology," 78.

## Chapter 13. Going to Extremes

1. Zapol, "Diving Physiology of the Weddell Seal," 1049.
2. Engel, *The Sea*, 47–49.
3. Kylstra, "Liquid Breathing and Artificial Gills," 156.
4. Clark and Gollan, "Survival of Mammals Breathing Organic Liquids," 1755–1756.
5. Kylstra, "Liquid Breathing," 159.
6. Ibid., 162.
7. Greenspan, et al. "Liquid Ventilation," 106–111.
8. Somero, "Environmental and Metabolic Effects of Pressure," 557–558.
9. Ibid., 560.

## Chapter 14. Air as Medicine

1. Sophocles, *Oedipus Tyrannus*, 97, 1012.
2. Hoff, *Bibliographic Sourcebook*, 303.
3. Cobb and Goldwhite, *Creations of Fire*, 190. Thomas Lovell Beddoes (1803–1849), a recently dethroned reader of chemistry at Oxford, started his Pneumatic Medical Institution with the belief that certain "factitious" gases could benefit humans suffering a variety of ills. Carbon dioxide enemas were used for "putrid" fevers. For tuberculosis, Beddoes advocated the inhalation of the gases derived from dung and had his patients overnight in the hayloft of local barns. Despite these questionable pursuits, Sir Humphrey Davy (1778–1829) was of such mind that he was able to conduct worthwhile experiments at the Institute to launch a career that brought him to discover potassium, barium, sodium, and earth alkalis. Davy was knighted in 1012.
4. Ibid., 192.
5. Junod, "Recherches Physiologiques," 353–354.
6. Hoff, *Bibliographic Sourcebook*, 304–305.
7. Triger, "Air Comprimé," 93–94. "A laborer who had been deaf ever since the battle of Antwerp heard better while in the compressed air than his other coworkers."
8. Hoff, *Bibliographic Sourcebook*, 306–307.
9. Milliet, "De l'Air Comprimé," 173–175.
10. Hoff, *Bibliographic Sourcebook*, 306.
11. Smith, *Effects of High Atmospheric Pressure*, 28, 31–32.
12. Hoff, *Bibliographic Sourcebook*, 307.
13. Kindwall, "Compressed Air Work," 5.
14. Tibbles and Edelsberg, "Hyperbaric Oxygen Therapy," 1642.

15. Ibid.
16. Kindwall, "Hyperbaric Oxygen's Effect on Radiation Necrosis," 477–479, and "Hyperbaric Oxygen Treatment of Radiation Cystitis," 591.
17. Tibbles and Edelsberg, "Hyperbaric Oxygen Therapy," 1644–1646. The 1996 Committee Report of the Undersea and Hyperbaric Medical Society approved the following conditions for hyperbaric oxygen therapy: Air or gas embolism, gas gangrene, decompression sickness, carbon monoxide poisoning, necrotizing soft tissue infections, osteomyelitis and bone infections resistant to antibiotics, radiation tissue injury; burns, brain abscesses, crushed limb injuries, and severe blood loss. Source: http://www.uhms.org/indic.htm, January 30, 1997.
18. Since Internet data and websites change frequently, the best method to obtain information from the hyperbaric oxygen and diving communities is to utilize a major search engine using "hyperbaric oxygen therapy" as the keyword phrase. Other indexed sites can be found at the Undersea and Hyperbaric Medical Society website at http://www.uhms.org and the Divers' Alert Network at http://www.dan.org.

Chapter 15. High Altitudes and Decompression Sickness

1. Fulton, "Historical Introduction," 2. Water boils when the kinetic energy of the water molecules is greater than the surface tension due to air pressure, and the kinetic energy of the water can be raised high enough if one heats the water—or lowers the air pressure. This latter phenomenon illustrates the ease of boiling water at higher altitudes. Conversely, higher pressures *raise* the boiling point of liquids. This is used commonly in pressure-cookers, where stews and casseroles are prepared at temperatures greater than those obtainable at 1 ATA. It is well known that in a vacuum, water at room temperature can "boil." There was concern among flight engineers, that in the relative vacuum of high altitudes and in space, gas tensions in the body would behave dangerously.
2. Fulton, memorandum, May 2, 1945.
3. Stever and Haggerty, *Flight*, 90–91.
4. Gaursaux, "Explosive Decompression," 164.
5. Fulton, "Recent Developments in Aviation Medicine," 263.
6. Stever and Haggerty, *Flight*, 90–91.
7. Guyton, "Whistling Death," appendix A.
8. Fulton, memorandum from Vannevar Bush, National Research Council, the Committee of Military Medicine, to Lewis Weed, the Medical Division chairman.

9. Henry, *New Haven Register*, June 8, 1941.
10. Fulton, *Diaries*, October 7, 1940.
11. Fulton, memorandum, May 2, 1945.
12. Fulton, preface, ix.
13. Nims, "Environmental Factors Affecting Decompression Sickness," 197.
14. Ibid., 198.
15. Harvey, "Physical Factors in Bubble Formation," 109–110.
16. Bateman, "Preoxygenation and Nitrogen Elimination," 270–272.
17. E. P. Kindwall and L. Holtgrewe, personal communications, 1996.
18. Kerut et al., "Detection of Right to Left Shunts in Divers," 371.
19. Moon et al., "Patient Foramen Orale and Decompression Sickness," 513.
20. Busuttili, "Hot from the U.S."

# Bibliography

Agricola, G. *De Re Metallica*, trans. Herbert and Lou Hoover. London: Mining Magazine, 1912.

Arlidge, J. T. *The Hygiene, Diseases, and Mortality of Occupations*. London: Percival, 1892.

Armstrong, H. G., and J. W. Heim. "The Effect of Flight on the Middle Ear." *Journal of the American Medical Association* 109 (1937): 417–421.

Arnold, J. H. "Liquid Breathing: Stretching the Technological Envelope." *Critical Care Medicine* 24 (1996): 4–5.

Association of Diving Contractors Therapy Guidelines, Committee on Science and Education. San Diego: Subsea International, 1994.

Bachrach, A. J. "A Short History of Man in the Sea." In: *The Physiology and Medicine of Diving and Compressed Air Work*, ed. P. B. Bennett and D. H. Elliott. Baltimore: Williams and Wilkins, 1975.

Barral, G. "Triger." In: *Le Panthéon Scientifique de la Tour Eiffel*. Paris: 1892.

Barre-Sinoussi, F., J. C. Chermann, et al. "Isolation of a T-Lymphotropic Retrovirus from a Patient at Risk for Acquired Immune Deficiency Syndrome (AIDS)." *Science* 20 (1983): 868–871.

Bassoe, P. "The Late Manifestations of Compressed Air Disease." *International Congress of Hygiene, 15th Congress* 3 (1912): 626–638.

Bassoe, P. "The Late Manifestations of Compressed Air Disease." *American Journal of Medical Science* 145 (1913): 526–542.

Bateman, "Preoxygenation and Nitrogen Elimination." In: *The Physiology and Medicine of Diving and Compressed Air Work*, ed. P. B. Bennett and D. H. Elliott. Baltimore: Williams and Wilkins, 1975, 270–272.

Behnke, A. R. "Decompression Sickness Following Exposure to High Pressures." In: *Decompression sickness*. Philadelphia: W. B. Saunders, 1951, 53–89.

———. "Effects of High Pressures; Prevention and Treatment of Compressed Air Illness." *Medical Clinics of North America* 26 (1942): 1213–1237.

Behnke, A. R., Jr., and J. P. Jones, Jr. "Preliminary BART Tunnel Results." *Dysbaric Osteonecrosis* 1 (1974): 25–34.

Behnke, A. R., and O. D. Yarbrough. "Physiologic Studies of Helium." *Naval Medical Bulletin* 36 (1938): 542–558.

Bennett, P. B., "The High Pressure Nervous Syndrome: Man." In: *The Physiology and Medicine of Diving and Compressed Air Work*, ed. P. B. Bennett and D. H. Elliott. Baltimore: Williams and Wilkins, 1975, 248–263.

———. "Inert Gas Narcosis." In: *The Physiology and Medicine of Diving and Compressed Air Work*, ed. P. B. Bennett and D. H. Elliott. Baltimore: Williams and Wilkins, 1975, 207–231.

Bennett, P. B., J. R. Blenkarn, and D. Youngblood. "Suppression of the High Pressure Nervous Syndrome in Human Deep Dives by He-N$_2$-O$_2$2." *Undersea Biomedical Research* 1 (1974): 221–237.

Bernal, J. D. *The Social Sciences in Science and History*. Cambridge: MIT Press, 1981.

Berner, R. A., and D. E. Canfield. "A New Model for Atmospheric Oxygen over Phanerozoic Time." *American Journal of Science* 289 (1989): 333–361.

Bert, P. *Barometric Pressure: Researches in Experimental Physiology*, trans. A. Hitchcock and L. Hitchcock. Columbus: College Book, 1943.

Beyerstein, G. "Why Another Therapy Scheme?" Unpublished guidelines, Subsea International, 1995.

Bichard, A. R., and H. J. Little. "Drugs That Increase Gamma Aminobutyric Acid Transmission Protect against the High Pressure Nervous Syndrome." *British Journal of Pharmacology* 76 (1982): 447–452.

Biographical Archives. "Andrew H. Smith, M.D." New York: New York Medical Society Library, ?1925, 175.

Birch, T. *The Works of the Honourable Robert Boyle*. 6 vol. London: J. and F. Rivington, 1772.

Black, A. *The Story of Tunnels*. New York: McGraw-Hill, 1937.

Boorstin, D. J. *The Discoverers*. New York: Random House, 1983.

Bordley, J., and A. M. Harvey. *Two Centuries of American Medicine*. Philadelphia: W. B. Saunders, 1976.

Bottazzi, F. *Leonardo Da Vinci*. New York: Reynal, 1988.

Bounhiol, J.-P. "Modification du Régime de Fixation de l'Oxygène Respiratoire Chez les Animaux Vivant en Milieux Suroxygenés." *Comptes Rendues de la Société Biologique de Paris* 101 (1929): 684–686.

Bowra, C. M. *Classical Greece*. New York: Time, 1966.

Boycott, A. E., and G. C. C. Damant. "Experiments on the Influence of Fatness on Susceptibility to Caisson Disease." *Journal of Hygiene* 8 (1908): 445–456.

Boycott, A. E., G. C. C. Damant, and J. S. Haldane. "Prevention of Compressed Air Illness." *Journal of Hygiene* 7 (1907): 343–425.

Bühlmann, A. A. "Decompression Theory: Swiss Practice." In: *The Physiology and Medicine of Diving and Compressed Air Work*, ed. P. B. Bennett and D. H. Elliott. Baltimore: Williams and Wilkins, 1975, 348–365.

Bullock, J., J. Boyle, M. B. Wang, and R. R. Ajello, eds. *Physiology.* New York: John Wiley and Sons, 1984.

Burchell, S. C. *Age of Progress in Great Ages of Man.* New York: Time-Life Books, 1966.

Burton, B. T. "Hazardous Materials." In: *Emergency Medicine: A Comprehensive Review*, 3rd ed., ed. T. C. Kravis et al. New York: Raven Press, 1993, 747–760.

Busuttili, M. "Hot from the U.S." http://www.divernet.com/usrep696.htm, 1996.

Cuilluux, A. "Notice sur la Vie et les Travaux de M. Triger." *Bulletin de la Société Géologique de France* 25 (1868): 547–559.

*Cambridge Biographical Dictionary*, ed. M. Magnusson and R. Goring. Cambridge: Cambridge University Press, 1991.

Carter, J. T. "The Role of Government in Preventing Ill Health at Work." In: *Hunter's Diseases of Occupations*, 8th ed., ed. P. A. B. Raffle, P. H. Adams, P. J. Baxter, and W. R. Lee. London: Edward Arnold, 1994, 15–20.

Catton, B. *The Civil War*, ed. R. M. Ketchum. New York: Doubleday, 1960.

Chinard, F. P. "Priestley and Lavoisier: Oxygen and Carbon Dioxide." In: *A History of Breathing Physiology*, ed. D. F. Proctor. New York: Marcel Dekker, 1995, 203–222.

Clark, L. C., and F. Gollan. "Survival of Mammals Breathing Organic Liquids Equilibrated with Oxygen at Atmospheric Pressure." *Science* 52 (1966): 1755–1756.

Clenney, T. L., and L. F. Lassen. "Recreational Scuba Diving Injuries." *American Family Physician* 53 (1996): 1761–1764.

Cobb, C., and H. Goldwhite. *Creations of Fire: Chemistry's Lively History from Alchemy to the Atomic Age.* New York: Plenum Press, 1995.

Curtis, S. E., J. T. Peek, and D. R. Kelly. "Partial Liquid Breathing with Perflubron Improves Arterial Oxygenation in Acute Canine Lung Injury." *Journal of Applied Physiology* 75 (1993): 2696–2702.

Dalton, John T. *Treatise on Human Physiology.* Philadelphia: Beacham and Lea, 1864.

Davis, R. H. "Deep Diving and Underwater Rescue." *Journal of the Royal Society of Arts* 82 (1934): 1032–1047.

*Dictionary of Scientific Biography*, ed. C. C. Gillispie. New York: Charles Scribner's Sons, 1972.

Diehl, L. *The Late, Great Pennsylvania Station.* New York: American Heritage Press, 1985.

Doll, R. E., Lt. "Decompression Sickness Among U.S. Navy Operational Divers: An Estimate of Incidence Using Air Decompression Tables." U.S. Navy Experimental Diving Unit, Research Report 4–64, NWP. Washington, D.C., 1965, 1–24.

Druitt, R. *The Principles and Practise of Modern Surgery.* Philadelphia: Blanchard and Lea, 1852.

DuBois, A. B., G. F. Bond, and K. E. Schaefer. "Alveolar Gas Exchange During Submarine Escape." *Journal of Applied Physiology* 18, no.2 (1963): 509–512.

Dugan, J., R. C. Cowen, B. Barada, L. Marden, and R. M. Crum, eds. *World Beneath the Sea.* Washington, D.C.: National Geographic. 1967.

Eads, J. B. "The Effects of Compressed Air on the Human Body." *Medical Times* 2 (1871): 291–293.

Ellenbog, U. *Von den Giftigen Besen, Tempfen und Reuchen.* Munich: Munich Press, 1927.

Empleton, B. E., E. H. Lanphier, J. E. Young, and L. G. Goff, eds. *The New Science of Skin and Scuba Diving.* New York: Association Press, 1974.

*Encyclopaedia Britannica.* 15th ed. Chicago: Encyclopaedia Britannica, 1992.

End, E. "The Use of New Equipment and Helium Gas in a World Record Dive." *Journal of Industrial Hygiene* 20 (1938): 511–520.

Engel, L. *The Sea.* New York: Time, 1961.

Faires, V. M. *Thermodynamics.* London: Collier-Macmillan, 1971.

Fitzgerald, R. S., "The Regulation of Breathing." In: *A History of Breathing Physiology*, ed. D. F. Proctor. New York: Marcel Dekker, 1995, 303–342.

Flexner, A. *Medical Education in the United States and Canada.* New York: Carnegie Foundation for the Advancement of Teaching, 1910.

François, M. "Des Effets sur les Ouvriers Travaillant dans les Caissons Servant de Base aux Piles du Pont du Grand Rhin." *Annales d'Hygiène Publique et de Médicine Légale* 14 (1860): 289–319.

French, G. R. W. "Diving Operations in Connection with the Salvage of the U.S.S. 'F-4.'" *Naval Medical Bulletin Washington* 10 (1916): 74–91.

Fulton, J. F. *Diaries*. John F. Fulton Collection, Yale Medical School Historical Library, New Haven.

———. Foreword. In *Barometric Pressure: Researches in Experimental Physiology* by Paul Bert, trans. A. Hitchcock and L. Hitchcock. Columbus: College Book, 1943.

———. "Historical Introduction." In: *Decompression Sickness*, ed. J. F. Fulton et al. Philadelphia: W. B. Saunders, 1951, 1–3.

———. *Papers*. John F. Fulton Collection, Yale Medical School Historical Library, New Haven.

———. "Physical Aviation and Medical Preparedness." *Connecticut State Medical Journal* 10 (1940, 23–42.

———. "Recent Developments in Aviation Medicine." *New England Journal of Medicine* 225 (1941): 263–268.

———. "Some Factors Affecting the Incidence of the Bends at Altitude." *Surgeon* 94 (1944.

Gauger, P. G., T. Pranikoff, R. J. Schreiner, F. W. Moler, and R. B. Hirschl. "Initial Experience with Partial Liquid Ventilation in Pediatric Patients with the Acute Respiratory Distress Syndrome." *Critical Care Medicine* 24 (1996) 16–22.

Gauroaux, P. P. "Explosive Decompression Ameliorated by Oxygen." *Bulletin de l'Académie de Medicine de Paris* 122 (1939): 164–167.

George, R. Y. "Functional Adaptations of Deep-Sea Organisms." In: *Functional Adaptations of Marine Organisms*, ed. F. J. Vernberg and W. B. Vernberg. New York: Academic Press, 1981.

Gersh, I., and H. R. Catchpole. "Decompression Sickness: Physical Factors and Pathologic Consequences." In: *Decompression Sickness*, ed. J. F. Fulton et al. Philadelphia: W. B. Saunders, 1951.

Gilbert, D. L. "Evolutionary Aspects of Atmospheric Oxygen and Organisms." In: *Environmental Physiology*. Philadelphia: W. B. Saunders, 1996.

Glossop, R. "Jules Triger, 1801–1867." *Geotechnique* 30 (1980): 538–539.

Gottlieb, L. S. *A History of Respiration*. Springfield, Ill.: C. C. Thomas, 1964.

Gould, S. J. *Wonderful Life: The Burgess Shale and the Nature of History.* New York: W. W. Norton, 1989.

Greenspan, J. S., M. R. Wolfson, S. D. Rubenstein, and T. H. Shafefer. "Liquid Ventilation of Human Preterm Neonates." *Fetal and Neonatal Medicine* 117 (1996): 106–111.

*Guinness Book of World Records,* ed. Peter Matthews. New York: Bantam Books, 1996.

Guyton, A. C. *Medical Physiology.* Philadelphia: W. B. Saunders, 1981.

Guyton, B. T. *Whistling Death.* Atglen: Schiffer Military/Aviation History, 1994.

Hackett, J. D. *Health Maintenance in Industry.* Chicago: A. W. Shaw, 1925.

Haldane, J. S., and J. G. Priestley. *Respiration.* New Haven: Yale University Press, 1935.

Hamilton-Paterson, J. *The Great Deep.* New York: Henry Holt, 1992.

Harvey, E. N. "Physical Factors in Bubble Formation." In: *Decompression Sickness,* ed. J. F. Fulton et al. Philadelphia: W. B. Saunders, 1951, 90–114.

Hatch, L. W., ed. "Labor Laws of 1909." *State of New York Department of Labor Bulletin* 41, ch. 291 (1909): 218–219.

Haxton, A. F., and H. E. White. "The Compressed Air Environment." In: *The Physiology and Medicine of Diving and Compressed Air Work,* ed. P. B. Bennett and D. H. Elliott. Baltimore: Williams and Wilkins, 1975.

Heller, R., W. Mager, and R. von Schrötter. *Luftdruckerkrankungen mit besonderer Berücksichtigung der Sogenannten Caissonkrankheit* (Vienna: Alfred Hölder, 1900.

Henry, T. R. *The New Haven Register.* New Haven, 1941.

Hewett, B. H. M., and S. Johannesson. *Shield and Compressed Air Tunnelling.* New York: McGraw-Hill, 1922.

Hills, B. A. *Decompression Sickness.* New York: Wiley, 1977.

Hoff, E. C. *Bibliographic Sourcebook of Compressed Air, Diving and Submarine Medicine.* Washington, D.C.: United States Navy, Bureau of Medicine and Surgery, 1948.

Hong, S. K., P. B. Bennett, K. Shiraki, Y.-C. Lin, and J. R. Claybaugh. "Mixed-Gas Saturation Diving." In: *Environmental Physiology.* Philadelphia: W. B. Saunders, 1996, 1023–1045.

Horstman, S. "Industrial Hygiene." In: *Public Health and Preventive*

*Medicine*, ed. J. M. Last and R. B. Wallace, 13th ed. Norwalk: Appleton and Lange, 1992.

Houghton, H. G. *Means of Supplying Air to Divers*. Hartford: John Russel, Jr., 1813.

Hudson, A., and Nelson, R. *University Physics*. San Diego: Harcourt Brace, 1982.

Jaminet, A. *Physical Effects of Compressed Air in the Construction of the Illinois and St. Louis Bridge etc.* St. Louis: R. and T. Ennis, 1871.

Jardine, F. M., and R. I. McCallum, eds. *Engineering and Health in Compressed Air Work, Proceedings of the International Conference.* London: E. and F. N. Spon, 1992.

Junod, V. T. "Recherches Physiologiques et Thérapeutiques sur les Effets de la Compression et de la Rarefaction de l'Air, tant sur le Corps que sur les Membres Isolés." *Revue Médicale Française et Etrangère; Journal des Progrès de la Médecine Hippocratique* 3 (1834): 350–368.

Kahlstrom, S. C., C. C. Butron, and D. B. Phemister. "Aseptic Necrosis of Bone. I. Infarction of Bones in Caisson Disease Resulting in Encapsulated and Calcified Areas in Diaphyses and in Arthritis Deformans." *Surgery, Gynecology and Obstetrics* 38 (1939): 129–146.

Keays, F. L. "Compressed Air Illness, with a Report of 3,692 Cases." *Publications of Cornell University Medical College* 2 (1909): 1–55.

Keenan, E. J. "Hormones of the Hypothalamus and Pituitary Gland." In: *Modern Pharmacology*. Boston: Little, Brown, 1986, 858–868.

Kerut, E. K., Truax, W. D., Borreson, T. E., Van Mieter, K. W., Given, M. B., and Giles, T. D. "Detection of Right to Left Shunts in Decompresoion Siokneoo in Divero." *American Journal of Cardiology* 79 (1997): 371–378.

Kindwall, E. P. "Compressed Air Work." In: *The Physiology and Medicine of Diving and Compressed Air Work*, ed. P. B. Bennett and D. H. Elliott, 4th ed. Philadelphia: W. B. Saunders, 1993, 1–18.

———. "Hyperbaric Oxygen's Effect on Radiation Necrosis." *Clinics in Plastic Surgery* 20 (1993): 473–483.

———. "Hyperbaric Oxygen Treatment of Radiation Cystitis." *Clinics in Plastic Surgery* 20 (1993): 589–592.

———. "Medical Aspects of Compressed Air Tunnelling: Background and Present State of the Art." In: *Tunnelling and Underground Space Technology*. Oxford: Pergamon Press, 1988, 169–173.

———. "Optimum Schedules for Caisson Decompression." In: *Engineering and Health in Compressed Air Work*, ed. F. M. Jardine and R. I. McCallum. London: E. and F. N. Spon, 1992, 157–171.

————. "A Short History of Decompression Sickness." In: *Diving Medicine*, ed. A. C. Bove, and J. C. Davis. Philadelphia: W. B. Saunders, 1990, 1–11.

————. "Treatment of Decompression Sickness: Where We Have Been and Where We Are Headed." Lecture presented at the Working Divers Conference, Little Creek, Ark., 1995, 1–20.

Kindwall, E. P., P. O. Edel, and H. E. Melton. "Safe Decompression Schedules for Caisson Workers." NIOSHA research grant. Milwaukee: Department of Hyperbaric Medicine, St. Luke's Hospital, 1983.

Kylstra, J. A. "Liquid Breathing and Artificial Gills." In: *The Physiology and Medicine of Diving and Compressed Air Work*, ed. P. B. Bennett and D. H. Elliott. Baltimore: Williams and Wilkins, 1975, 155–165.

Kylstra, J. A., R. Nantz, J. Crowe, W. Wagner, and H. A. Saltzman. "Hydraulic compression of mice to 166 Atmospheres." *Science* 158 (1967): 793–794.

Langlois, J.-P. "La Prophylaxie des Accidents dans l'Air Comprimé." *Revue Générale des Sciences Pures et Appliquées* (1911) 22: 54–60.

Lapp, R. E. *Matter*. New York: Time, 1963.

Lee, H. C., K. C. Niu, S. H. Chen, L. P. Chang, K. L. Huang, J. D. Tsai, and L. S. Chen. "Therapeutic Effects of Different Tables on Type II Decompression Sickness." *Journal of Hyperbaric Medicine* 6 (1991): 11–17.

Legge, T. "Arsenic Poisoning." In: *Industrial Health*, ed. G. M. Kober and E. R. Hayhurst. Philadelphia: P. Blakiston's Son, 1924.

Levy, E. "Compressed-Air Illness and Its Engineering Importance, with a Report of Cases at the East River Tunnels." *Technical Papers Bureau of Mines. Washington* 285 (1922): 1–46.

————. "Workers in Compressed Air: Precautions Adopted by the N.Y. Public Service Commission for Protecting Their Health." *Scientific American* (supplement) 84 (1917): 73–74.

Lundgren, C. E. G., A. Harabin, P. B. Bennett, H. D. Van Liew, and E. D. Thalmann. "Gas Physiology in Diving." In: *Environmental Physiology*. Philadelphia: W. B. Saunders, 1996.

McCullough, D. *The Great Bridge*. New York: Simon and Schuster, 1982.

MacDonald, A. G. "Hydrostatic Pressure Physiology." In: *The Physiology and Medicine of Diving and Compressed Air Work*, ed. P. B. Bennett and D. H. Elliott. Baltimore: Williams and Wilkins, 1975, 78–102.

Macinnis, J. A. "Open-Sea Diving Techniques." In: *The Physiology and Medicine of Diving and Compressed Air Work*, ed. P. B. Bennett and D. H. Elliott. Baltimore: Williams and Wilkins, 1975, 20–33.

McLennan, J. C. "Helium: Its Production and Uses." *Nature* 105 (1920): 778–780.

Magnus, O. *Historia om de Nordiska Folken* (History of the Nordic People). Stockholm: Gidlund, 1976.

Malizieux, M. "Fondations à l'Air Comprimé." *Annales des Ponts et Chaussies* 5 (1874): 329–402.

Masur, H., et al. "An Outbreak of Community-Acquired Pneumocystis carinii Pneumonia." *New England Journal of Medicine* 305 (1981): 1431–1438.

Mayow, J. *Medico-Physical Works, Being a Translation of Tractatus Quinque Medico-Physici*, trans. A. C. Brown. Edinburgh: Alembic Club, 1907.

Middleton, W. E. *The History of the Barometer*. Baltimore: Johns Hopkins University Press, 1964.

Miller, S. L. "A Production of Amino Acids Under Possible Primitive Earth Conditions." *Science* 117 (1953): 528–529.

Milliet, J. "De l'Air Comprimé au Point de Vue Physiologique." *Gazette Médicale de Lyon* 8 (1856): 172–177.

Moir, E. W. "Tunnelling by Compressed Air." *Journal of the Royal Society of Arts* 44 (1895–1896), 567–583.

Moon, R. E., E. M. Carpenter, and J. A. Kisslo. "Patent Foramen Orale and Decompression Sickness in Divers." *Lancet* 1 (1989): 513–514.

Moon, R. E., R. D. Vann, and P. B. Bennett. "The Physiology of Decompression Sickness." *Scientific American* 269 (1995): 71–77.

Nims, L. F. "Environmental Factors Affecting Decompression Sickness." In: *Decompression Sickness*, ed. J. F. Fulton et al. Philadelphia: W. B. Saunders, 1951, 192–241.

Parker, R. *Miasma: Pollution and Purification in Early Greek Religion*. Oxford: Clarendon Press, 1983.

Piantadosi, C. A., and E. D. Thalmann. "Thermal Response of Humans Exposed to Cold Hyperbaric Helium-Oxygen." *Journal of American Physiology: Respiration, Environment, Exercise, and Physiology* 49 (1980): 1099–1117.

Pol, B., and T. J. J. Watelle. "Méthode sur les Effets de la Compression de l'Air, etc." *Annales d'Hygiène Publique et de Médecine Legale (Industrielle et Sociale)* (1854): 241–279.

Primm, J. N. *Lion of the Valley*. Boulder, Colo.: Pruett, 1990.

Proctor, D. F. "History of Breathing Physiology." In: *Lung Biology in Health and Disease*, ed. D. F. Proctor. Cambridge: Cambridge University Press, 1986.

Ramazzini, B. *De Morbus Artificium Diatriba* (1700), trans. W. C. Wright. Chicago: University of Chicago Press, 1940.

Resnick, R., and D. Halliday. *Physics for Students of Science and Engineering*. New York: John Wiley and Sons, 1963.

Ringqvist, T. "The Ventilatory Capacity in Healthy Subjects." *Scandinavian Journal of Clinical Laboratory Investigation* 18 (1966): 1–179.

Rosen, G. *The History of Public Health*. New York: MD Publications, 1958.

Shapin, S., and S. Schaffer. *Leviathan and the Air-Pump: Hobbes, Boyle and the Experimental Life*. Princeton: Princeton University Press, 1985.

Siegel, F. P., et al. "Severe Acquired Immunodeficiency in Male Homosexuals, Manifested by Chronic Perianal Ulcerative Herpes Simplex Lesions." *New England Journal of Medicine* 305 (1981): 1439–1444.

Sigerist, H. E. *A History of Medicine*. Vol. 2, *Early Greek, Hindu, and Persian Medicine*. Oxford: Oxford University Press, 1961.

Singstad, O. "Industrial Operations in Compressed Air." *Journal of Industrial Hygiene* 18 (1936): 497–523.

———. "The Queens Midtown Tunnel. Discussion." *Proceedings of the American Society of Civil Engineers* 70 (1944): 375–386.

Smith, A. H. *Effects of High Atmospheric Pressure Including the Caisson Disease*. Brooklyn: Eagle Print, Brooklyn, 1873.

Sodeman, W. A., Jr., and T. M. Sodeman. *Pathologic Physiology*. Philadelphia: W. B. Saunders, 1979.

Somero, G. N. "Environmental and Metabolic Effects of Pressure." *Annual Review of Physiology* (1992): 557–577.

Sonntag, R., and G. Van Wylen. *Introduction to Thermodynamics, Classical and Standard*. New York: John Wiley and Sons, 1982.

State of New York. *Annual Report of the Industrial Commissioner— 1922, New York State Department of Labor*. No. 28. Albany: J. B. Lyon, 1923.

———. Industrial Code Bulletins, "Work in Compressed Air." Nos. 22 and 22-A. Albany: Department of Labor Board of Standards and Appeals, 1922, 7–10.

———. New York Labor Bulletin, "Labor Laws of 1909." Albany:

New York State Department of Labor, 1910, 218–221.

Stever, H. G., and J. J. Haggerty. *Flight.* New York: Time, 1965.

Stigler, R. "Die Kraft Underer Inspirations Muskulatur." *Pflügers Archiv für die gesamte Physiologie des Menschen un der Tiere* 139 (1911): 234–254.

Temple, R. *The Genius of China; 3,000 Years of Science, Discovery and Invention.* New York: Simon and Schuster, 1986.

Thackrah, C. T. *The Effects of Arts, Trades, and Professions on Health and Longevity.* Canton, Mass.: Science History Publications, 1985.

Thomson, E. "Helium in Deep Diving." *Science* 65 (1927): 36–38.

Tibbles, P. M., and J. S. Edelsberg. "Hyperbaric Oxygen Therapy." *New England Journal of Medicine* 334 (1996): 1642–1647.

Triger, C.-J. "Application de l'Air Comprimé pour le Sauvetage des Batimants." *Compte Rendu Hebdomadaire des Séances de l'Académie des Sciences* 21 (1845): 233–234.

———. "Mémoire Sur un appareil de l'Air Comprimé Pour le Percement des Pults, etc." *Compte Rendu Hebdomadaire des Séances de l'Académie des Sciences* 13 (1841): 90–94.

———. *Letters.* C.-J. T. to L'Académie des Sciences, Institut de France, Pochette de Séance, April 14, 1834.

———. *Minutes.* Assemblée Générale, Centre Historique Minier, Paris: 1841–1847.

Undersea Medical Society (UMS). "Study to Determine the Feasibility for Developing Interim Decompression Schedules for Tunnel Workers." Workshop report to NIOSHA, prepared by UMS, July 25, 1979.

U.S. Department of Health and Human Services. "Kaposi's Sarcoma and Pneumocystis Pneumonia among Homosexual Men—New York City and California." *Morbidity and Mortality Weekly Reports* 30 (1981): 305–307.

———. "Pneumocystis pneumonia—Los Angeles." *Morbidity and Mortality Weekly Reports* 30 (1981): 250–252.

*United States Navy Diving Manual,* Part 1. Washington, D.C.: United States Navy, 1959.

*United States Navy Diving Manual,* Part 1. Washington, D.C.: United States Navy, 1970.

Urey, H. C. "On the Early Chemical History of the Earth and the Origin of Life." *Proceedings of the National Academy of Sciences* 38 (1952): 351–363.

Vann, R. D., J. Grimstad, and C. H. Nielsen. "Evidence for Gas Nuclei in Decompressed Rats." *Undersea Biomedical Research* 7 (1980): 107–112.

Vernon, H. M., "The Solubility of Air in Fats and Its Relation to Caisson Disease." *Lancet* 2 (1907): 691–693.

Vivenot, R. R., Jr. "Historischer Rückblick auf die Entwicklung der Aërotherapie." *Allgemeint Wiener Medizinische Zeitung* 15 (1870): 9–10, 21–22.

Walder, D. N. "The Prevention of Decompression Sickness." In: *The Physiology and Medicine of Diving and Compressed Air Work*, ed. P. B. Bennett and D. H. Elliott. Baltimore: Williams and Wilkins, 1975, 456–466.

Ward, C. A., D. McCullough, and W. D. Fraser. "Relation Between Complement Activation and Susceptibility to Decompression Sickness." *Journal of Applied Physiology: Respiration, Environment, Exercise and Physiology* 62 (1987): 1160–1166.

Ward, C. A., D. McCullough, D. Yee, D. Stanga, and W. D. Raser. "Complement Activation Involvement in Decompression Sickness of Rabbits." *Undersea Biomedical Research* 17 (1990): 51–66.

Ward, G. *The Civil War*. New York: Alfred Knopf, 1990.

Weindling, P. *The Social History of Occupational Health*. London: Croom Helm, 1985.

West, G. *Innovation and the Rise of the Tunnelling Industry*. Cambridge: Cambridge University Press, 1988.

Whittaker, A. H., and D. J. Sobin. "Ulrich Ellenbog." In: *Historical Milestones in Occupational Medicine and Surgery, Industrial Medicine* (1941): 1–12.

Williams, C. T. "The Compressed Air Bath and Its Uses in the Treatment of Disease." *British Medical Journal* 1 (1885): 769–772, 824–828, 936–939.

Wilson, M. *American Science and Invention*. New York: Simon and Schuster, 1954.

Woodward, C. M. *A History of the St. Louis Bridge*. St. Louis: G. I. Jones and Company, 1881.

Yodaiken, R. Memorandum. "OSHA Decompression Tables." Office of Occupational Medicine, Occupational Safety and Health Administration, Washington, D.C., August 2, 1988.

Zapol, W. M. "Diving Physiology of the Weddell Seal." In: *Environmental Physiology*. Philadelphia: W. B. Saunders, 1996, 1049–1056.

# Index

Diving: bells, development of,
39–45, 160; breath-hold, 160;
closed system, 160–62; helmet,
161–65; SCUBA, 163–64, 172;
towers, U.S. Navy, 175; "dry"
(chamber), 183
—equipment: Renaissance, 38;
Roman, 38
—suits: Siebe, 160, 162–63;
Neufeldt-Kuhnke (1930), 161
Duffner, G. J., 155
Duffner tables. *See* Decompression
tables; Washington State tables
Dysbaric osteonecrosis, 152–56

Eads, J. B., 61–62, 71, 136
Edel, Peter, 156–57
End, Edgar: early helium dives
and, 169
Experimental Diving Unit (EDU),
169, 181

Fat: absorption of nitrogen gas by,
118–21, 124, 127
Fish: adaption to extreme depths by,
110, 194, 229*n*1, effects of water
pressure on, 110
Fleuss, Henry, 164
Flier's bends. *See* High-altitude
bends
François, M.: and early caisson's
disease, 53, 57
Fructus, Xavier, 176
Fulton, J. F., 207–08
FX-80: use of, in liquid breathing,
190–91

Gagnan, Emil, 164, 172
Galen, 11

Gas: word coined, 10; Boyle's Law
of, 20, 42; physics of, 21; Dalton's
Law of, 27, 119; Henry's Law of,
27, 119, 127, 230*n*18; and 19th-
century methods of analysis, 116,
117; dangers of, at increased
pressure, 167; noble, 167
Gollan, F.: liquid breathing and,
190–91
Goodman, M. W., 174
Greathead, J. H., 99
Guericke, Otto von, 17, 29, 221*n*2

Hahnemann, S. C., 228*n*21
Haldane, J. B. S., 126
Haldane, J. S., 121–25; experiments
of, 124–25
Haldane effect, 123
Haldanian decompression. *See*
Decompression tables, staged
Halley, Edmund, 42
Harvey, William, 11–12
Haskin, D. C., 97, 100–01
Helium: use of, in diving, 168–69,
171, 182; safety of, 169; as cause
of bends, 182; effects of, 185
Henderson, Yandell, 204–05
Henry, James, 119
Henry's Law, 127, 230*n*18
Henshaw, Thomas, 198
High-altitude bends (flier's bends),
204–06, 210
High-pressure nervous syndrome
(HPNS), 183–84
Hippocrates, 8
Holland, C. M., 151
Homeopathy, 228*n*21
*Homo sapiens:* limitations of, to
atmospheric pressure, 189

Hooke, Robert, 18–20, 22, 24
Horsepower: term coined, 33
Hydreliox (hydrogen-helium-oxygen), 185
Hydrogen: use of, in diving, 185, 234n35
Hyperbaric medicine: first studies of, 89–90
Hyperbaric oxygen therapy (HBO), 201–03; indications for, 236n16

Jacobs, George, 106
Jaminet, Alphonse, 71–74, 93
Japp, Henry, 142
JIM (diving suit, late 20th century), 162

Keays, F. L., 142
Kindwall, Eric, 155–57, 178, 188, 230n21, 234n41
Kylstra, J. A., 190–91

Labor laws, 139–40
Laboratory method: growth of, 93, 113, 116
Lambert, Albert, 233n5
Lavoisier, A. L., 25–26
Lee, H. C., 177
Legge, Sir Thomas, 138–39
Le Prieur, Yves, 164
Letheridge, John, 161
Leucippus, 8
Levy, Edward, 143–44, 151
Link, Edwin, 188
Liquid breathing, 190–93
Liquids: physics of, compared to gases, 22; boiling points of, 236n1; effect of pressure on, 236n1
Lundgren, Claus, 181

Lung: functional structure of, 5; Galenic notions of, 11; early theories of function, 11–12; forces on, 38–39, 222n14; and response to injury, 192

Mal de caisson, 48, 52. See also Decompression sickness
Mayow, John, 13, 220n16
Medical lock: medical recompression and, 91; first use of, 103; Moir's, 108; physicians and, 141; first proposal of, by A. H. Smith (1870s), 227n17
Medicine: growth of, 53
Miasma, 7, 197, 219n3
Microbubbles: formation of, after decompression, 130
Mixed gas diving, 169, 184–85
Moir, E. W., 95–96, 100, 102–05, 229n31

New York City: commerce of (1860–70s), 95, 226n1
New York State: establishment of regulations in, 143, 147; Labor Commission, 145–47
— compressed air regulations: of 1909, 232n26; of 1921, 147; of 1922, 148
Newcomen, Thomas, 33
Newton, Isaac, 9, 219n8
Nitrogen: historical source of, 5; discovery of, 25; as causative gas of the bends, 102, 123; naming of, by Lavoisier (1774), 116; first isolation of, in bends victims, 117; absorption of, by fatty tissue, 118–20; isolation of, in

nervous tissues (Haldane), 125;
elimination of, by lungs, 157–58,
174; narcotic effect of, 167–68
Nitrogen-helium-oxygen (trimix),
185
Nitrox (nitrogen-oxygen) decompression, 176
Niu, K. C., 177
Nohl, M. G., 169

Occupational health. *See* Worker
safety
Occupational Safety and Health
Administration (OSHA), 93, 140;
shortcomings of, 155–56, 159
One-atmosphere diving, 160, 161
Oxygen: toxic effect of, 4, 167–68,
174; discovery of, 23, 25; as
treatment for gas poisoning,
121; use in decompression, 158,
174–75

Pascal, Blaise, 17
Phlogiston, 25
Physicians: company, 137, 141
Pilot tube technique: in tunnel-
ing, 102
Pneumatic therapy, 198–99, 235n3
Pol, B., 52–55
Pressure: water, 17, 22, 39, 45–46,
131, 166, 193–95, 222n14; high-
altitude, 205–06; units of, 220n5;
vacuum, force of, 221n2, 221n3
Priestley, Joseph: and discovery of
oxygen, 25
Priestly, J. G.: and work with John
Haldane (1900s), 122
Proteins: effect of high pressure on,
193–95

Ramazzini, Bernardino, 138
Recompression chamber (medical
lock): as treatment for the bends,
56, 91, 92, 103; first design, 91;
first use of, 103–04; benefits of,
108; as treatment for limb pain,
125; by Sir Robert Davis, 130–32
Roebling, J. A., 77–79
Roebling, Washington, 78, 80–81,
89, 92, 136
Rouquayrol, Benoit, 163–64

Sandhogs: attitudes of, 71, 231n10,
232n23; hardships of, 143, 149
Saturation diving, 187; world rec-
ord, 234n41
Savery, Thomas, 31
Schaffer, Thomas, 192
Scientific thought: advances in, 93,
110–12
SCUBA diving: development of,
163–64, 172
Shield technique: for tunneling, 99
Siebe, Augustus, 161
Smeaton, John, 42–45
Smith, A. H., 86–87; theories of,
88, 90, 93; and criticism of theo-
ries of Paul Bert, 118–19
Split-shift decompression, 141,
143, 154
Staged decompression, 166
—New York State regulations of
*1909*, 147, 232n26; of *1921*, 147;
of *1922*, 148
Steam engine: development of, 32–35
Stilson, G. D., 165
Sturmius, J. C., 42
Supersaturation, 230n13: of tissue,
theory of, 127–28